THE AESTHETIC BRAIN

THE AESTHETIC BRAIN

How We Evolved to Desire Beauty and Enjoy Art

Anjan Chatterjee

OXFORD
UNIVERSITY PRESS

OXFORD
UNIVERSITY PRESS

Oxford University Press is a department of the University of Oxford.
It furthers the University's objective of excellence in research, scholarship,
and education by publishing worldwide.

Oxford New York
Auckland Cape Town Dar es Salaam Hong Kong Karachi
Kuala Lumpur Madrid Melbourne Mexico City Nairobi
New Delhi Shanghai Taipei Toronto

With offices in
Argentina Austria Brazil Chile Czech Republic France Greece
Guatemala Hungary Italy Japan Poland Portugal Singapore
South Korea Switzerland Thailand Turkey Ukraine Vietnam

Published in the United States of America by
Oxford University Press
198 Madison Avenue, New York, NY 10016

First issued as an Oxford University Press paperback, 2015.

Library of Congress Cataloging-in-Publication Data
Chatterjee, Anjan.
The aesthetic brain: how we evolved to desire beauty and enjoy art/Anjan Chatterjee.
pages cm
Includes bibliographical references and index.
ISBN 978-0-19-981180-9 (hardback); 978-0-19-026201-3 (paperback)
1. Aesthetics—Psychological aspects.
2. Genetic psychology. 3. Brain—Evolution. I. Title.
BH301.P78C45 2013
111′.85—dc23 2013014485

To my father and my mother

CONTENTS

PART TWO
PLEASURE

PART THREE
ART

PREFACE

Like the proverbial fool, in writing about aesthetics, I may be rushing into places where angels fear to tread. Why should a scientist write about aesthetics? Art and aesthetics have long resided deep within the humanities. Philosophers, historians, critics, and artists themselves have had much to say. Could viewing aesthetics through the lens of science possibly contribute to the vast knowledge and deep insights that humanists have gathered over the years? At first glance it would appear unlikely.

Even if it is doubtful that science has something worthwhile to say about aesthetics, scientists like me might feel optimistic. This optimism, perhaps born out of ignorance, is bolstered by two central ideas. The first idea is that all human behavior, at least at the level of the individual, has a neural counterpart. There is no thought, no desire, no emotion, no dream, no flight of fancy that is not tethered to the activity of our nervous system. And so the scientist supposes that a fine understanding of the properties of the brain will shine its own special light on any human faculty. Such faculties include language, emotion, and perception, and there is no reason to expect that this understanding could not extend to aesthetics. The second idea is that evolutionary forces have sculpted our brains and behavior. Evolutionary biology and more recently evolutionary psychology provide a powerful framework from which to consider the forces that shape why we do what we do. Ideas from neuroscience and evolutionary psychology have not been historically available to scholars in the humanities. They have not been combined with each other to inform discussions about art and aesthetics. Thus, scientists who think that neuroscience will tell us the "how" of aesthetics and that evolutionary psychology will tell us the "why" of aesthetics might well be indulged our optimism. Perhaps we scientists have something worthwhile to say about aesthetics after all.

Some humanists hear this kind of bravado and remain skeptical. This skepticism takes many forms. Some simply dismiss the idea of scientific aesthetics as irrelevant. Others might consider it yet another fad destined to bask in its 15 minutes before fading back into proper obscurity. Still others are frankly hostile to the idea. Perhaps such reactions arise from general anxieties that neuroscientific barbarians are invading everywhere. Just in the last decade terms like neuroeconomics, neurolaw, neuroliterature, and even neurotheology have infested our language. Now even such a sacrosanct domain as aesthetics is not safe from neurohordes. Will neuroaestheticians plunder aesthetics of its riches and leave it a desiccated hull of its true glory?

The neuroscientist might be tempted to disregard such humanist reactions as the territorial cries of an endangered species. The humanities are under assault when it comes to jobs and funding and in some quarters are in danger of being left behind as science and technology marches on. The neuroscientist might dismiss humanist approaches to aesthetics as a nostalgic nod to the past. This is a past to which one might politely pay homage, while believing that the future lies ahead with science and technology. Or the neuroscientist might view the humanist response as anti-intellectual. Why reject any form of inquiry? Surely in the marketplace of ideas, if neuroscience and evolutionary psychology have anything to contribute, these contributions will survive. So the neuroscientist might think, let's forge ahead, do what we do, and not worry about humanist concerns.

I have drawn caricatures of neuroscientists and humanists. The reality is that most neuroscientists care little about the humanities, at least in their professional lives, and I suspect that scholars in the humanities care little about science as it relates to their fields of expertise. Of those who have more than a passing interest in the other domain of inquiry, many are open to considering points of convergence or are at least open to the possibility of a meaningful conversation. For examples of books that seek such convergences, take a look at Margaret Livingstone's *Vision and Art: The Biology of Seeing* and John Onian's *Neuroarthistory*.

I think that it is premature to be optimistic or pessimistic about the contributions of neuroscience and evolution to aesthetics. Neuroaesthetics is a growing field. It is still working out its research agenda, methods of investigation, and even which questions are worth pursuing. It is also too early to know how best to think of evolutionary psychology in the context of aesthetics. Again, the appropriate methods and what might be regarded as convincing evidence for a point of view are debated within the field. These early days in the science of aesthetics make it an exciting time. Horizons are wide open. New discoveries seem possible.

There are, however, inherent tensions between the humanities and science that we should acknowledge. The ways that these tensions will play out as neuroaesthetics evolves remain to be seen. At the outset it is worth being explicit about the tensions, so that we can keep an eye on them even as we press on. Some tensions might be resolved along the way and others might well prove to be insurmountable. Three main tensions lie in the cracks between the humanities and the sciences. These tensions are between subjective and objective experiences, between concerns with the particular and the general, and between expansive and reductive approaches to aesthetics and art.

Have you ever had the experience of being so absorbed by a painting or a piece of music that you lose all sense of space and time? These magical moments in which we lose our sense of self are ironically deeply subjective. The problem, of course, is that science demands some objectivity in its analysis. How does one capture this profound experience objectively? Typically, objectivity takes quantitative form. Translating these seemingly transcendent experiences into numbers is critical for an experimental approach to aesthetics. Information needs to be quantified, hypotheses need to be tested, and claims need to be replicated or falsified. These are the basic foundations upon which progress in science is built. Will such objectification, which is the stuff of science, completely miss its mark when it comes to describing aesthetic experiences? Will neuroaesthetics be blind to the central essence of these experiences and be relegated to piddling at its margins?

Aesthetic experiences, one might argue, grow out of a deep involvement with the particular. I might adore visual art, you might prefer music, and someone else dance. And within visual art, I might be awestruck by Cezanne and you by Pollock. And within Cezanne, I might delight in his landscapes and still lifes, but his portraits might leave me cold. Perhaps the nature of an aesthetic experience is to be captivated by very particular objects. Any generalization that can be made from that experience is trivial. For example, the generalization that I like Cezanne because of the way he depicts contours and uses planes to convey volume is just not all that important to my experience of Cezanne's paintings. The goal of science is to uncover generalizations. The particular is simply the vehicle used to uncover a general principle. As a matter of experimental design, we usually include many particulars to make sure that any annoying and unique contribution by an individual piece of art (or stimulus as we are apt to call them) is counterbalanced by equally annoying and unique contributions made by other pieces of art. Whatever is left, then, is something that is common to all the pieces under consideration. The more pieces used, the greater the power to get to the generalization we seek. This particular–general tension is a counterpart to the subjective–objective tension that resides in the person observing or creating art. Here the tension lies in the work of art itself. Is the scientist's desire to generalize across different works of art again missing the magic that can only be found when deeply engaged with particular works?

Any scientific approach to a complex domain must sort the domain into its component parts and then examine each part in relative isolation. As scientists come to understand the parts better, a view of the whole system emerges. This approach has been very successful in advancing our understanding of perception, language, emotions, actions, and, as I shall discuss later, how we make decisions. But is this approach, which is necessary if there is to be a science of aesthetics, a recipe for failure? Perhaps the aesthetic experience is an emergent property of different components, which cannot be derived by studying its parts. The situation might be like a chemist studying the properties of hydrogen and oxygen with the goal of understanding water. Once again, the aim of aesthetic science might be off its true mark.

So, where are we? In what follows, I touch on traditional ideas in aesthetics, while keeping an eye on these tensions between science and the humanities. This book has sections on beauty and pleasure and art. Beauty is integral to how most people think about aesthetics. What makes something beautiful? What emotions does beauty evoke? Aesthetic experiences can be profoundly emotional and are often pleasurable. How is pleasure, and more generally our experiences of rewards, organized in the brain? Why do we have specific pleasures? Beauty and pleasure are complicated enough topics that they can and have been discussed independently. My goal here is to connect them to each other and to art. I will explore how we might think about art and its relationship to beauty and to pleasure. In a postmodern world, has art divorced itself from its traditional partners, beauty and pleasure? I will gaze at beauty, pleasure, and art through the bifocal spectacle of neuroscience and evolutionary psychology.

Let me acknowledge some notable omissions in the book. I do not consider nonvisual art, such as music and literature. I also do not talk about the applied art of architecture. Some of the ideas in this book will be relevant to these domains, but drawing out relevant connections would take us too far afield from what is already a meandering journey. I have also set aside the important topic of creativity. Creativity could easily be the basis for another book. My emphasis is on aesthetic encounters, not the creation of conditions for such encounters. Finally, I will spare you the many caveats and the tentative nature of conclusions that characterizes scientific writing. In the process, I hope I do not do too much violence to what we know, or how we know, or the nature of scientific knowledge and its accretion.

As we proceed, I confess to feeling a bit optimistic about connecting aesthetics to neuroscience and evolutionary psychology. A focus on the brain will help us understand the *how* of aesthetics, and frameworks from evolutionary psychology will help us understand the *why* of aesthetics. But I am getting ahead of myself. Let's see if neuroscience and evolutionary psychology can light our way in this labyrinthine world of beauty, pleasure, and art.

INTRODUCTION

On a sunny afternoon, I walked to the Museum of Modern and Contemporary Art in Palma, Mallorca. The museum is nestled between ancient walls and set behind a wide patio overlooking the vibrant blue Mediterranean. Before walking into the museum, I paused at this patio and gazed at the waving palm trees, bobbing boats, and beautiful light glistening on the bay.

In the museum, I first delighted at ceramic plates made by Picasso. They highlighted his mastery at rendering faces with an economy of lines. Moving into a room with prints made by Miró, I was entranced by his ability to use gestures so expressively. Getting lost in the artwork of these two masters gave me great pleasure.

I then wandered into the main exhibit that was called "Eros y Thanatos" (Love and Death). Off to one side, I noticed a series of plastic bags filled with what looked like blood. They were hanging off tubing as if to be used in transfusions. The bags were set in two spiraling circles one above the other. I walked around the bags staring at them, simultaneously curious and disturbed. Beyond these curtains of blood was a set of mirrors about 2 or 3 feet high set at angles to each other. They reflected the legs of visitors, including my own. Past these mirrors was a small tree with a narrow trunk hanging upside down. On the branches of the tree sat many small dark birds, aligned right-side up. And on the floor were more of these birds, as if they had fallen off like little ripe fruit. I found myself intrigued by these displays, even as I was bewildered by what they meant.

A museum guard noticed my bewilderment and started to speak to me in Spanish. Seeing that his overture only increased my confusion, he switched to English. Kindly, he directed me to some laminated sheets in a corner of the room that described what each artist was doing with their displays. I needed this decoder to help interpret what I was seeing. The decoder had nothing to do with being an American in Spain. I could have been in the Institute of Contemporary Art on the campus of the University of Pennsylvania where I work. Even there, I need explanations to engage with exhibits on display.

This anecdote raises fundamental questions that I will try to address in this book. When I stood looking out over the Mediterranean, I found the light, the water, and the colors of the scene beautiful. But what made this a beautiful sight? I am quite sure that my parents, if they had been standing there with me, would have found the view beautiful. I suspect that you would too. Is something special happening in our brains when we look at these beautiful scenes? If my intuition that most people would find the view beautiful is correct, then is there something universal about my response to its beauty? Besides environmental scenes like the one before me, faces and bodies can also be particularly beautiful. Is it just a coincidence that these things, environmental scenes, portraits, and nudes, are prominent subjects in art and that they also have their special locations in different parts of the brain? Does something similar happen in my brain when I experience the beauty of faces and bodies as when I experience the beauty of scenes?

Discovering what happens in the brain when we are captivated by beautiful objects won't completely solve the question of why the objects are beautiful. To address the why question, we turn to evolutionary psychology. The basic idea of evolutionary psychology is that our mental abilities, like our physical traits, evolved if they enhanced our survival. Our ancestors from a distant past adapted certain behavioral traits to survive tough environments and to choose partners that would give them healthy children. When it comes to beauty in people, certain physical features of faces and bodies advertised a person's health. These features, which were important in choosing a mate tens of thousands of years ago, are what we now regard as beautiful. When it comes to beauty in scenes, some places were more inviting to our hunter-gatherer ancestors wandering around in the distant past. These scenes looked both safe and rich in resources that would help small bands survive a life that was nasty, brutish, and a long time ago.

When looking out over the Mediterranean, a calm pleasure came over me. I felt as if I could rest a while and soak in this scene, much like my hunter-gatherer ancestors might have. However, the pleasure seemed different then the pleasure of admiring an attractive woman on the patio. It also seemed different than the pleasure of eating paella with its rich and complex and fragrant tastes. What are these pleasures and how do they relate to aesthetics or to art? Looking at the Picasso plates and the Miró prints also gave me pleasure. Here I found myself delighted at their mastery at the same time I enjoyed the vibrant colors and lines. I thought one of the Miró prints would look especially nice in my home and wondered how much such a print would cost.

These pleasurable experiences raise the question of what it means to have an aesthetic experience. In the brain, our emotional systems for pleasure are housed in deep structures far from the surface. These structures have imposing names like the orbitofrontal cortex and the nucleus accumbens. We also have various brain neurotransmitters like opiates, cannabinoids, and dopamine that do the signaling work for our pleasures. But are different pleasures exchanged into a single currency in the brain? Is the pleasure of the scent of a magnolia in anyway similar to that of winning a bet or to admiring a Rothko painting?

In trying to relate our pleasures to aesthetic experiences, we run into a paradox. The paradox is that we evolved our responses to beauty because they were useful for survival, and yet, aesthetic responses to beauty are not supposed to be useful. Specific reward systems in the brain are tied to our appetites. So beautiful faces and bodies are tied to our sexual desires and landscapes are tied to our desire for safety. Adaptive traits are adaptive precisely because they were useful for the most practical of reasons: find a mate, make healthy babies, and survive a cruel world. However, eighteenth-century theoreticians like Anthony Ashley-Cooper (the third Earl of Shaftsbury) and Immanuel Kant thought that aesthetic encounters are a special class of experiences involving "disinterested interest."

The pleasure of an aesthetic experience is self-contained and does not extend beyond itself into somehow becoming useful. When I looked at the Miró print and thought that it would look spectacular in my home, that fantasy may have been pleasurable (probably not as pleasurable as actually buying the print). However, the pleasure of acquisition is not aesthetic. If this idea is correct and aesthetic experiences are disinterested (a view that is by no means universally accepted), then we arrive at the paradox. How can disinterested interests be adaptive? Analogously, if the rewards are adaptive, meaning that they are useful, how can they be aesthetic? As I meander through the world of beauty, pleasure, and art in this book, I shall address this paradox.

What would disinterested interest actually mean in the brain? To get a handle on this, I will turn to experiments that use pleasure seen in faces of rats. From a series of clever experiments, the neuroscientist Kent Berridge and his colleagues have identified two parallel reward systems that he calls liking and wanting. These systems are close to each other in the brain and usually work in concert. We like what we want and we want what we like. Even though these systems normally work together, they can part ways. What would liking without wanting be? It would be a pleasure without an acquisitive impulse. Perhaps this is what it means to have aesthetic pleasure, liking without being contaminated by wanting. What would wanting without liking be? The classic example is drug addiction. Addicts crave their drugs well past the point of liking them. Addiction is the prototypic antiaesthetic state.

So far, in relating the ideas of beauty and pleasure to my experience in the museum at Palma, I have not touched upon the "Love and Death" exhibit (spiraling blood, angled mirrors, fallen birds) that I found bewildering. While the art pieces were interesting, I did not find them beautiful. I did not appreciate them in the way that I enjoyed Picasso and Miró. The pieces evoked various emotions in me, but these emotions were not pleasant. I felt some mixture of intrigue, annoyance, confusion, and even disgust. Were these pieces really art? Most of us associate art with beauty. We usually expect to experience pleasure when encountering art. How do we reconcile these examples of contemporary art with many people's ideas of art as beautiful and pleasurable? Is there now a big divide, in which artists, art critics, and curators are the high priests of culture who need to explain the mysteries of art to the rest of us?

Blake Gopnik, who was for many years an art critic at the *Washington Post,* suggests that much of art is part of an ongoing cultural conversation in which the meaning, or the various messages that an artwork can convey, is paramount. These messages rarely have anything to do with beauty and pleasure. Like a complicated conversation, it is unreasonable to expect that we could naively step in the middle of its stream and expect to understand what is going on. Gopnik thinks that scientists investigating art are preoccupied with aesthetic theories of the eighteenth century. Neuroaesthetics and evolutionary aesthetics may be cutting-edge science, but they are old-fashioned thinking and are simply out of touch with art as it is practiced today.

My general response to these concerns in relating scientific aesthetics to art is that our goal is to provide a framework that is broad enough to accommodate all art: contemporary or art of any other period. It should even be broad enough to accommodate art made 50 or 200 years into the future. To accomplish this goal, we recognize a critical triad lies at the core of art encounters and aesthetic experiences. This triad, as pointed out by the psychologist Art Shimamura [1], consists of sensations, emotions, and meaning. Sensations in art might be vivid colors and bold lines when it comes to painting, or meter and melody when it comes to music. Emotions evoked by art are often pleasure, but they can also be disgust or any number of other subtle emotions. Meaning in art might be political, intellectual, religious, ritualistic, or subversive. It might involve a cultural conversation in which the artist engages the viewer. We need a framework that incorporates the details of sensations, encompasses various emotions, and makes contact with a range of possible meanings. Different people engaged with art, like art critics or art historians, or philosophers of art, might very well emphasize any one part of the triad. Whichever turns out to be important to works of art or to expert theoreticians, we have a scientific framework from which to test ideas. However, as we shall see, contemporary art does pose a special challenge for scientists. The challenge lies in figuring out if science can deal with meaning in art and whether this challenge sets inherent limits on the reach of science.

Does the brain have a neural network dedicated to aesthetics? The evidence from brain studies shows that there is no specific neural network dedicated to aesthetics. Furthermore, if we break aesthetics down into component parts, it is evident that no specific neural networks are dedicated to aesthetic sensation or aesthetic emotion or aesthetic meaning. We shall see that art is organized in flexible ensembles. The notion of a flexible ensemble applies to both the brain and evolutionary mechanisms. We have ensembles of neural subsystems that combine flexibly to give aesthetic experiences. The specific subsystems that make up this aesthetic ensemble are guided by experience and culture. Since components of the ensemble can be engaged flexibly, what is regarded as art or what evokes an aesthetic experience can change over time and differ across places. For example, impressionist paintings are now adored by the general public but were initially regarded with disdain. Of course, the human brain has not changed from the late nineteenth century to the early twenty-first century. Our neural perceptual machinery and reward systems remain the same. The change is in the link between specific percepts and reward systems based on our knowledge and experience. This flexibility by which components combine in aesthetic ensembles is part of what makes art and aesthetic experiences rich and even unpredictable.

A final question to which I will turn to in this book is: do we have an art instinct? An instinct is a behavioral adaptation that evolved because it gave our ancestors an advantage in dealing with their environments. As I examine the evidence for an art instinct, it will become quite clear that determining what constitutes adaptive behavior, especially when it comes to something as complex as art, is not a simple task. We are forced to imagine what it might have been like for our ancestors in particular time periods over many years and in vastly different environments. Our current living conditions, the institutions that surround us, and the things that make individuals successful in a modern society are for the most part not relevant to why we evolved to be the way we are. We humans are creatures of the past.

The pieces of ourselves that evolved to help us survive in the past are just that, pieces. Not all pieces that make up our minds are adaptations. Some pieces came along for the ride. The paleontologist Steven Jay Gould referred to evolutionary by-products as "spandrels." A spandrel is the triangular architectural space that is formed by columns and arches that meet in a room. The spandrel has no functional purpose in the architectural design of the building. It is simply a by-product of other features that have structural importance, like the columns, arches, and walls. But rather that remaining in the background, these spaces can be highlighted. They can be decorated as is often done in churches and other classical buildings. Spandrels are accidents, in the sense that these spaces are not a part of the structural design of the room. Yet they can be put to interesting uses and even be a focal point in the room.

Scholars who talk about the evolution of art tend to take one of two positions. They think of art as either an instinct or as an evolutionary by-product. The art-as-instinct advocates point out that art is everywhere. Whether we look back at ancient cultures or across widely differing contemporary cultures, we find examples of art. Surely if it is so pervasive, art is an instinct. The late philosopher Dennis Dutton went so far as to name his popular book *The Art Instinct*. The art-as-by-product advocates emphasize the idea of "art for art's sake" and that art practices are incredibly diverse and shaped by culture. The idea that art is for art's sake means that art is not useful for anything. It must be a by-product of other abilities that were useful to our ancestors. The sheer diversity of art even casts doubt on the idea that art is one thing at all. How could so many seemingly different objects be the expression of an instinct? I will show that neither the idea of art as an instinct nor the idea of art as an evolutionary by-product is very satisfying. We need a third way to think about art in order to understand both its universality and its diversity. A little bird bred in Japan over the last 250 years might point to this third way of thinking about art.

My walk began with gazing out at the beautiful Bay of Palma, moved on to admiring the elegance of Picasso and Miró and reflecting on my emotional responses, and ended with struggling through the bewildering display "Love and Death." In this book, I shall take this walk again, only more slowly. I shall begin with the excitement of beauty. I shall then meander through the delight of pleasure and end in the wonder of art. I will show you how our ancestors were driven instinctually to desire beauty and how we have since relaxed into enjoying art.

THE AESTHETIC BRAIN

PART I

BEAUTY

Chapter 1

What Is This Thing Called Beauty?

> Remember that the most beautiful things in the world are the most useless, peacocks and lilies for instance.
>
> —John Ruskin

> When you have only two pennies left in the world, buy a loaf of bread with one, and a lily with the other.
>
> —Chinese Proverb

Beauty has a hold on us. We are drawn to it. We go to great lengths to get it. We immerse ourselves in it. Beauty delights us. It inspires us. It makes us ache. It sends us into despair. If myths are to be believed, it can launch a thousand ships to war. But what exactly is this thing called beauty?

Is beauty a property of things out there in the world? Or is beauty to be found in our heads? Is beauty a fiction constructed by culture? Perhaps beauty is created by people of influence, who use beauty to maintain their own power or to make money by selling things. These ideas of beauty are at odds with a common view that beauty is fundamentally useless. There is also a long-standing belief dating back to the ancient Greeks that beauty is a core value, along with truth and goodness, that grounds our humanity. Thoughtful people have characterized beauty in one or another of these different ways. So, beauty is this powerful and mysterious thing that we crave. But we don't know where it is to be found and what it is all about.

Let us return to the first question. Is beauty out there in the world? Are objects beautiful? It seems silly to even ask the question. Any love-struck adolescent would think that the question was idiotic. Of course objects are beautiful! Faces, bodies, landscapes can all be beautiful. Their beauty has been enshrined by artists over the ages. Music and poetry can be beautiful. Perfumes and treasured meals are beautiful. Even mathematical proofs can be beautiful. It would seem obvious with all these examples that many objects in the world have great beauty.

These examples of beauty that seem to unequivocally establish that beautiful objects exist also reveal the problem. The objects are so different from each other that it is hard to grasp what makes them all beautiful. What do extraordinarily beautiful things like Ingrid Bergman's face, the Bryce Canyon at dawn, and the mathematical theorem called Euler's identity have in common? Is it a trick of language that we call these things beautiful? If beauty is contained in objects and we generally know objects through our senses, can objects that evoke no sensations in common really all be regarded as beautiful? The beauty of mathematics doesn't even arise from sensations. Maybe beauty is not in the object itself but is in something that is happening inside us. Maybe these objects are only beautiful in our heads and work by stirring beauty receptors in our brains. Maybe only special people have these receptors, and these special people with a refined taste for beauty need to explain beauty to the rest of us.

This question of whether beauty is in the world or within us has been tossed back and forth over the ages. It is a question that ultimately collapses on itself. The question presupposes that the world of objects and the perceiver of objects are separate entities. We have to choose between beauty being in one or the other. One of the lessons of evolutionary psychology, as we shall see in greater detail later, is that we are deeply integrated with the natural world. Our mind has been sculpted by nature and it is tightly coupled to the environment. We cannot ask questions about the structure of our minds without bumping into properties of the world. The question of whether beauty lies in the world or in our heads might be reframed as follows: what in the coupling of mind and world gives us the experience of beauty?

To examine this special coupling of mind and world, we will consider different objects that can be beautiful. We start with a discussion of faces. Scientists know quite a bit about the psychology and neurology of face processing. Bodies also can be beautiful. Many principles that we shall extract by examining beauty in faces will apply to beauty in bodies. We then turn to landscapes. Landscapes obviously differ from human faces and bodies. Are the aesthetic experiences of looking at beautiful people in any way similar to the experience of being in a beautiful environment? What can we learn about aesthetics by examining these objects that have preoccupied so many artists over the centuries?

Faces, bodies, and landscapes are all sensuous examples of beauty. We can revel in beautiful curves or soft light or lush colors when we absorb these kinds of beauty. What do we do with beauty that has no obvious sensations, like the beauty of an abstract idea? To explore this rarified beauty, we will peek at mathematics. Surely, if mathematics can be beautiful, its beauty must be very different from that of a sexy body. Along the way, we shall also keep an eye on how culture might influence our ideas and experiences of beauty.

Examining this thing called beauty that is both powerful and useless will prepare us for later discussions of pleasure and art. For many people, beauty is an essential ingredient of art. What is the relationship of beauty and pleasure? What is the relationship of beauty and art? Before we get to those questions, let's see what we can discover by exploring beauty in people, places, and proofs.

Chapter 2

Captivating Faces

The *Mona Lisa* is the most famous painting in the world. Her face grabs the viewer and has been the source of endless discussion. Her face is enigmatic. She is expressing something, but exactly what, seems obscure. Her face is but a striking example of the more general fact that faces engage us. When beautiful, faces captivate us. In a 2010 survey, Audrey Hepburn was judged to have the most beautiful face of the twentieth century [2]. As far as Hollywood icons are concerned, my own tastes lean toward Ingrid Bergman (didn't make the top 10) or Grace Kelly (ranked number 5). Different Web sites debate whether Cary Grant or Paul Newman was the most beautiful man of the twentieth century. Are we pawns of the Hollywood machine that manipulates us into believing that these people's faces are beautiful?

How do we decide if media manipulations, or other cultural contrivances, brainwash us into accepting sometimes impossible standards of beauty? Can responses to beauty be disentangled from the way the media and our culture more generally mold our tastes? Two research strategies help address these questions. The first is to see if people, especially when they are from different cultures, share opinions about beauty. The second is to see if babies, before their awareness has been shaped by culture, respond to beauty in the same way as adults.

The evidence from both these research strategies suggests that humans are pretty much hard-wired to regard some faces as more beautiful than others. Even though adults are inundated with media images of what is or is not beautiful, the results of many studies suggest that a core of what we all regard as beautiful exists independent of our cultural baggage. People of the same ethnic group within a culture rate the attractiveness of faces very similarly, even when the faces are of people from different ethnicities.

For example, men agree on which women of Asian, Hispanic, European, or African heritage look attractive [3]. By itself, that finding might be expected and could be consistent with the idea that our impressions of facial beauty are guided by common cultural influences. Beyond the judgments made by people within ethnic groups, opinions about facial beauty by people across ethnic groups are quite consistent. This observation might also be the result of common cultural influences affecting people despite their ethnic differences. However, studies that examine judgments of faces made by people across cultures show at least as high correlations of facial beauty as the correlations of judgments made by people of different groups within a culture [4]! Taken together, the consistency of ratings of facial attractiveness across ethnicities and cultures suggests that there are common elements in faces to which adults are responding that are not just a cultural creation. It is likely that cultural fashions often exaggerate and exploit biases toward beauty that most of us already have embedded within us.

The idea that facial beauty has universal underpinnings does not mean that culture does not affect which faces are regarded as attractive or that individuals are not biased by their own personal experiences. For example, some facial adornments in African and Southern American tribes are not regarded as beautiful by many with Western urban tastes. Closer to home, preferences for hairstyles and facial accessories like glasses and piercings and makeup can be very different from person to person.

Sometimes, even when differences in attractiveness judgments vary across cultures, the reasons for these differences might be general. In one study [5], Ache (from Paraguay) and Hiwi (from Venezuela) natives and people from the United States and Russia were shown faces of people from Brazil, the United States, and the Ache tribe. At the time that they were tested, these native tribes were quite isolated from the rest of the world. Everybody preferred women's faces with larger eyes and delicate jaws. These features, as we shall see later, are associated with youthfulness. Beyond that common trend, the Ache and Hiwi Indians did not agree very much with people from the United States and Russia. One could guess that culture must have an effect on what people find attractive, which is why their judgments differed. However, the Ache and Hiwi Indians were also isolated from each other. Despite having no contact with each other, the Ache and Hiwi natives agreed with each other in their beauty judgments. The overall physiognomies of Ache and Hiwi features are similar. One explanation of the findings is that sheer exposure to facial characteristics is how, in this case, environmental experiences contributed to the respective differences in attractiveness judgments. The Ache and Hiwi Indians agreed with each other because they had similar experiences with different kinds of faces.

What about babies? Faces enthrall babies. Within an hour of being born, babies notice images that look like faces [6]. In their first week they distinguish their mother's face from other faces and they mimic facial expressions that they see [7]. Babies hold a gaze directed at them and often smile in response to faces. While it seems clear that babies orient to faces, how do we know if they find one face more attractive than another? Even though babies don't talk the way we might sometimes want them to, their behaviors give us a pretty good idea about what they like.

Developmental psychologists use a technique of preferential looking to determine what attracts babies' attention. These psychologists present babies with two faces side by side and then measure how long a baby looks at either face. From these measurements, they can tell which face is more attractive to the baby in the sense of literally attracting babies' eyes [8]. Judith Langlois and her colleagues showed that infants as young as a few days old and certainly by 3 months of age gaze longer at faces that adults find attractive. These attractive faces could be faces of men or women, those of other babies, and even faces of people from different races [9]. Babies' preferences for attractive faces can also be shown in other ways. In one experiment, 1-year-old babies were given dolls that were identical except for having attractive or unattractive faces. The babies played with the dolls with attractive faces almost twice as long as the dolls with unattractive faces [10]. It appears that babies respond to facial attractiveness in a way that cuts across age, race, and gender. Importantly, they respond in this way before they have been influenced by Hollywood, Bollywood, Estée Lauder, or *People* magazine.

When we talk about attractiveness in faces, we usually mean sexual attractiveness. But, "attractiveness," or the power of objects to attract our attention, is influenced by the context in which we see these objects. The cognitive psychologist Helmut Leder, with his colleagues, showed students pictures of people in the streets of Vienna [11]. The students were given two stories. In one story they were told that Vienna is a great place if you are young and single. It has an active social scene, and it is a place where one easily meets potential partners. The other story emphasized that Vienna is a big city, and like most big cities, it has crime. In the active social-scene story, students looked longer at the faces of men and women that were sexually attractive. In the dangerous big-city story, the bias to look at attractive women's faces did not change. However, attractive men's faces no longer drew people's attention. Since street violence is associated with men more than with women, in the dangerous-city context men were more salient as potential threats than as potential sexual partners. Physical beauty was no longer relevant to the experience of looking at men. My point is that our responses to faces are influenced by the context in which we encounter them. As we shall see later, context has a profound influence on how we experience most objects and, importantly, how we experience pleasure and art.

To summarize, the evidence at hand suggests that some faces are universally regarded as attractive. This claim does not deny cultural and contextual influences on attractiveness, or that individuals may develop different tastes. However, these relative influences are typically built on a universal bedrock of attractiveness. The idea of universal features of beauty in faces brings up three questions that we shall explore. First, if beautiful faces have universal features, can these features be measured? Measurement is after all the life-blood of any science. Second, if they are universal, are they "hard-wired"? Colloquially, the term *hard-wired* implies that our response to these features is built into our brains in the same way. Finally, if they are universal, why did we evolve to find these features attractive?

Chapter 3

The Measure of Facial Beauty

As a child, growing up in India, I remember hearing the following story. When God decided to make humans, He molded dough into human forms and put them in the oven. When God pulled out the forms, He saw that they looked white and pasty and He wasn't happy. So God threw the batch out and started again. This time He left the tray in the oven too long. The forms came out black and burnt. So He threw those out and tried again. Finally, He got it right. Out came the golden brown color perfect for humans. The story drives home the point that most groups make up stories to proclaim their own exceptionality. The earliest attempts to measure beauty, especially with faces, were contaminated with such biases, despite claiming to be "objective."

The long history of European attempts to determine the physical features that make a face beautiful unsurprisingly ended up claiming that white European features were the most attractive. For example, a statue of Apollo (Belvedere), dated to 320 BCE and discovered near Rome around 1496 CE, was regarded as the epitome of beauty. For 400 years after its discovery, it was the most famous sculpture in the Western world. Beauty was a matter of figuring out how well facial features matched this and other such icons of beauty from antiquity. The eighteenth-century Dutch artist and anatomist Petrus Camper measured facial angles in profiles. The angle was derived using one line from the ear to the lip and another from the forehead to the most protruding part of the jaw, often the upper lip. Camper found that Greek statues had a profile angle of about 100 degrees. Most human profile angles range from 90 to 70 degrees. Using these measurements, Camper claimed that beauty in races improved in the order of African to East Asian to White European features [12], with the European profile being the closest to the ideal established by Greek statues.

We all have a tendency to want to relate physical features to character. The Swiss pastor Johann Casper Lavater, in his 1772 *Essays on Physiognomy,* wrote confidently that the chin signifies strength in a man [13]. He claimed that an angular or receding chin is seldom found in "discreet, well disposed, firm men." He also asserted that horizontal eyebrows that are "rich and clear always convey understanding, coldness of heart, and the capacity to frame plans." Perhaps not surprisingly, he thought that European facial physiognomy was superior to others. In a curious connection, FitzRoy, the commander of the ship *Beagle,* was a Lavater fan. The *Beagle* is the ship that took Darwin around the world to gather evidence from which he later developed his theory of evolution. Darwin wrote in his biography that the captain doubted that anyone with Darwin's nose had "sufficient energy and determination for the voyage." Of course, this was the nose that led Darwin through 5 years of arduous travel that inspired his theory of evolution.

The idea that facial and other physical features were indicative of personality continued into the early twentieth century. A physician, Katherine Blackford, promoted a "science" of character analysis based on physical features. Her books [14], which went through several editions, urged American businesses to use this science in what became known as the "Blackford plan." In referring to skin color, she asserted "always and everywhere the normal blond has positive, dynamic, driving, aggressive, domineering, impatient, active, quick, hopeful, speculative, changeable and variety loving characteristics; while the normal brunette has negative, static, conservative, imitative, submissive, cautious, painstaking, plodding, slow, deliberate, serious, thoughtful, and specializing characteristics." These examples show that early attempts to objectify beauty and characterize personalities from physical features were often exercises in prejudice masquerading as science.

Setting aside parochial prejudices, can facial attractiveness be measured reliably? Three parameters contribute to facial attractiveness, none of which is unique to any specific ethnicity. The first parameter is averageness. The second is symmetry. Both of these parameters apply to men and women. The third parameter has to do with features that make men and women look different from each other, or the parameter of sexual dimorphism.

Averageness as a measure of attractiveness was discovered serendipitously. Before Katherine Blackford deemed blondes positive, Francis Galton was interested in whether specific facial features were characteristic of personality traits. Galton, Charles Darwin's cousin, was a brilliant statistician, anthropologist, and explorer. He invented statistical correlations and fingerprinting. He also promoted eugenics. He became interested in whether one could identify common features in the faces of criminals. He overlaid the faces of many criminals "convicted of murder, manslaughter, or robbery accompanied with violence" onto a single photographic plate, hoping that the composite face would reveal the prototype look of a criminal. Instead of discovering the criminal mastermind, Galton found that composite faces were more attractive than each individual face that made up the composite [15]!

Galton discovered that averaged facial features are attractive. Modern research methods confirm this unexpected discovery [16]. We should be clear that an "averaged" face is not the same as a plain face. These faces have statistically averaged features, such as how thick or thin a nose is, or how far apart the eyes are set. Earlier, there was doubt about the validity of averaging experiments. The concern was that composite faces blurred the edges of each individual face, making them look younger. They had the soft-focus haze often used by fashion photographers. However, recent computer techniques have avoided this methodological limitation and it is clear that faces representing the central tendency of a group are seen as more attractive than individual faces. Even infants look at these "averaged" faces longer than they look at other faces [17].

Another quantitative parameter of faces that people find attractive is symmetry. The anthropologist Karl Grammer and the biologist Randy Thornhill measured facial symmetry by measuring the distance of different facial landmarks on both the left and right side of the geometric center of the face. They showed that this symmetry index correlated with judgments of attractiveness of men's and women's faces [18]. Many subsequent experiments have confirmed these results. One interesting study was able to hone in on the effects of symmetry by using pictures of identical twins, who of course look very similar [19]. But twins have subtle facial differences; even if their genes are identical, their environmental exposures are not. The investigators first established which of the two twins had a more symmetrical face. They found that the more symmetrical twin was also regarded as more attractive. Thus, in these pairs of faces, which are similar in so many ways, symmetry influenced their attractiveness.

Sexual dimorphism refers to differences in physical features based on gender. We saw that averaging and symmetry have similar effects on attractiveness for both men's and women's faces. But, what differentiates their attractiveness? The sex hormones, estrogen and testosterone, produce sexually dimorphic physical features. Estrogen feminizes and testosterone masculinizes features. Heterosexual men, regardless of their culture, find feminized features in women attractive [20].

The physical effects of estrogen are similar to what we see in babies' faces. Faces that are baby-like have large eyes, thin eyebrows, big foreheads, round cheeks, full lips, small noses, and small chins. People just like these "cute" features. This fact was not lost on Walt Disney. In 1928, Mickey Mouse made his animated appearance in a movie called *Steamboat Willie*. Mickey started out long and lithe. In 1935, his animator gave him a pear-shaped body, added pupils, and shortened his nose. The curious case of Mickey Mouse is that he has become more like a baby, with a bigger head and bigger eyes and smaller limbs, even as he got older over the last 80 years.

Pictures of adult men's and women's faces can be artificially made to look more or less baby-like. Does this manipulation affect attractiveness? Men tend to find women who look younger than their chronological age and have some baby-like qualities more attractive. Men prefer women with high foreheads, big eyes, small noses, full lips, and small chins. These features, associated with high levels of estrogen, signal fertility in women. However, men do not find one baby feature, big puffy cheeks, attractive in adult women. Men like high cheekbones, which is a sign of maturity. Men, it appears, prefer features that signal both youthfulness and fertility, but with an added dash of sexual maturity [21].

When talking of averageness in women's features, one point needs to be made. Averaged faces are very attractive, but they are not off the charts. Averaged faces often win beauty pageants, but they are not the faces of supermodels that grace the covers of most fashion magazines. The psychologist David Perret showed that composites of the best-looking women are more attractive than composites of an entire group [22].

Supermodel-level attractive women have exaggerated rather than averaged features. They have larger eyes, thinner jaws, and smaller distances between their mouths and chins than average. These are exaggerated versions of features that distinguish women's from men's faces. The faces of supermodels often have features that are typical of young girls, sometimes of girls under 10 years of age!

The story of what heterosexual women find attractive in men is even more complicated. Across culture after culture, in listing what they find attractive, women rank physical attractiveness less highly than men do. Women are not driven by visual cues as much as men are. The computational neuroscientists Ogi Ogas and Sai Gaddam, in their entertaining book *A Billion Wicked Thoughts*, marshal considerable evidence to make this point. They use data from what they call "the world's largest behavioral experiment to reexamine one of the most important and intimate of all behaviors: sexual desire." They analyzed Internet searches to find out what men and women chose to search for on the Web. When it comes to desires in the virtual world, gender differences are strikingly clear. Men overwhelmingly search for pornography. The videos are visually graphic without much in the way of plot or emotional engagement. By contrast, women overwhelmingly search for e-Rom Web sites. These sites tell romantic stories often built around a heroic man. Women's desires are formed by many different signals, besides what a man looks like. Status, power, wealth, the ability to protect and provide are more important to women than to men. Henry Kissinger, a man not known for great physical beauty, was often accompanied by young, very attractive women. He observed, "Power is the ultimate aphrodisiac."

Although women are more complicated than men in terms of who they think is attractive, they do respond to specific male physical features. Testosterone gives a face a bigger, squared-off jaw, thin cheeks, and a heavy brow. In general, women prefer these "masculinized" faces, a preference that is widespread across cultures. Even among the remote !Kung San, bushmen with broader chins and more robust bodies end up with more sexual partners [23].

However, women find masculinized features attractive only to a point. If men's faces are too masculine, woman experience them as domineering. This impression that men with broad chins are domineering has been seen in people in many cultures and might not be far from the truth. West Point cadets with more masculinized features end up higher in the military hierarchy while at school and later in their careers compared to their somewhat feminized-looking classmates [24]. If women want stable partners to help raise their children, then a man who is too domineering might not be the best choice for a long-term partner. He might not be invested in the relationship or in the family. So, it turns out that women prefer faces with masculine features and a little femininity thrown in [25]. Slightly feminizing masculine faces cuts the domineering note and makes men seem warm, emotionally available, and likely to be committed to the relationship.

Another fascinating subtlety about what women find attractive in a man is that their preferences vary during their menstrual cycle. This change in preference is called the "ovulatory shift hypothesis" and turns out to be a robust finding in human attractiveness research [26]. Young women find different men attractive depending on whether they want a short-term or a long-term partner. When considering a short-term partner, women want more masculine-looking men. This preference is exaggerated just before women ovulate and are most likely to get pregnant. By contrast, women's preferences for long-term male partners do not vary through the menstrual cycle. We shall come back to the implications of this ovulatory shift in women's preferences for short-term partners when we talk in more detail about the evolutionary reasons driving our preferences for attractiveness.

To summarize the findings regarding facial beauty, infants and adults, as well as people from different cultures, respond similarly to the same measurement parameters. Whether you are a man, a woman, or an infant, you probably find averaged and symmetrical faces attractive. Features that distinguish men from women are also attractive when the differences are emphasized. The context in which we see people makes a difference in how attractive we find them. Context effects have a powerful influence on our pleasures, as we shall see later. For women, the context of whether a man is found attractive can be power and status. When women orient to a man's physicality, the context can be whether she is looking for a fling or for someone to settle down with. That context also varies depending on whether or not she is close to ovulating.

When we think of people as being attractive, we also think of their bodies. In most cultures, we do not get to see naked bodies as often as we see exposed faces. But, if the principles of attraction in faces have biological and evolutionary underpinnings, as we shall explore later, we would expect similar principles to apply to the parameters that make human bodies attractive.

Chapter 4

The Body Beautiful

In his book *The Nude: A Study in Ideal Form,* Kenneth Clark pointed out that every time we criticize a human figure, for example, that the neck is too short or the feet too big, we reveal that we have an ideal of physical beauty. Clark's observations suggest that we can measure beauty in bodies. There is a long history of people trying to do exactly that.

In the second century, the Greek physician Galen argued that an arm that was three hand-lengths long was more beautiful than one that was two and a half or one that was three and a half hand-lengths long. The idea that beauty in the human body was a matter of proper proportions really took off in the European Renaissance. After studying in Italy, the German painter and mathematician Albrecht Dürer introduced laws of proportion into northern Europe. He described a system of ideal human proportions in his 1582 book, *De Symmetria*. His system reduced the body to simple forms, such as cylinders, spheres, cones, cubes, and pyramids, that could be measured easily. He constructed a proportional system that was actually based on his own hand. The middle finger was supposed be equal to the width of the palm, and the width of the hand was supposed to be proportionate to the forearm. From a set of relations of fingers to hand, hand to forearm, forearm to arm, and limbs to height, he constructed a canon for the entire body. He thought that this system of identifying parts in relation to the total body length gave the body a harmonious, organic unity.

Scientific attempts to measure beauty in bodies have not been as extensive as studies of beauty in faces. However, some of the principles that make a face beautiful also apply to bodies. Symmetry is an important feature for both men's and women's bodies. Sexual dimorphic features in bodies when exaggerated are also attractive. As I mentioned in the last chapter, we have less exposure to bodies than we do to faces. Whether averaging bodies also makes them more attractive is not known. Given that we form averages by seeing many examples, and in most contemporary cultures we do not get to see many unclothed bodies (as compared to faces), we probably don't form the same kind of prototypes of averaged bodies.

Except for little dogs wearing cute coats when walking the streets of big cities, animals are inclined to show their bodies uncovered. Animals turn out to be very aware of each other's bodies. Remarkably, animals find symmetry of body parts attractive [27]. For example, male reindeer do well in their sexual marketplaces if their huge antlers are symmetrical [28]. Female swallows mate more often with males that sport large and symmetrical tails [29]. Body symmetry also affects beauty in the human animal. Men with symmetrical feet, ankles, hands, elbows, wrists, and ears are considered more attractive than lop-sided men [30]. The point is not that women necessarily fetishize these parts of men's bodies, but that these parts can be measured easily and are good markers of overall symmetry. Men with symmetrical bodies also do well in their own sexual marketplace. They tend to have sex a few years earlier than other men. They also have sex earlier when courting a specific woman, and have two or three times as many partners than less symmetrical men. Their partners even experience them as better in bed! It turns out that a man's physical symmetry can predict the likelihood of his female lover having an orgasm better than his earnings, investment in the relationship, or frequency of love-making [31].

Heterosexual men also prefer symmetrical women. This preference is evident in laboratory experiments as well as from behavioral observations. Physically symmetrical women have more sexual partners than less symmetrical women. It turns out that women with large and symmetrical breasts are more fertile than women with less symmetrical breasts. Women also become more symmetrical during ovulation. Symmetry in soft tissue as measured in women's ears and third, fourth, and fifth fingers can increase up to 30 percent during ovulation [32].

We saw that sexual dimorphic features can drive attractiveness in male and female faces. Sexual dimorphic features also influence how animals and people react to bodies [21, 27]. In the animal world, males are often extravagant in their displays. Exaggeration of plumage, as in the peacock, is common among birds. Female swordfish prefer males with longer swords, female swallows prefer males with longer tails, female flies of the family *Diopsidae* prefer males with long stems on their eyes, rutting reindeer prefer males with large antlers. Size matters.

Most physical differences between men and women are the result of the hormones testosterone and estrogen. Testosterone, among other things, increases human physical size. Most people like tall men [33]. The taller man almost always wins U.S. presidential elections. The CEOs of successful companies are more likely to be tall than to be short. Height can affect starting salaries. The link between height and status goes both ways. People thought to be powerful are seen as a few inches taller than if they were thought to be relatively powerless. Women find tall men attractive. Almost without exception, women prefer men that are of average height or above to men that are shorter than average. Tall men get more responses to personal advertisements. To be very concrete about sexual selection, when choosing sperm donors in fertility clinics, women are more likely to want the sperm of tall men [34].

Most people think that the ideal shape of a man's torso is the V shape, with broad shoulders and narrow hips [35]. Both men and women dislike a pear-shaped man, one with narrow shoulders and a wide waist. The major difference in strength between men and women is in the arms, chest, and shoulders. This is where testosterone has a big impact in laying down muscle mass. Not surprisingly, men's fashions over the years have emphasized and exaggerated their shoulders, from the use of epaulets that designate rank, to shoulder pads in just about every power suit on Wall Street. Romans wore breast-plates that emphasized the size of their chests. Now men insert pectoral implants and use liposuction to remove fat from their waist and breasts. Male models are stereotyped in their physique. They are over 6 feet tall with chest sizes of 40–42 inches and waists 30–32 inches. In male bodybuilders, these proportions are exaggerated, with chests almost twice as large as their waists.

Women have a different distribution of fat than men. Estrogen deposits fat in breasts, buttocks, and thighs. These women's body parts preoccupy men. In my lab, when we were designing an experiment on facial attractiveness, we first went to a Web site called "Hot or Not" to see if photographs from this site could be used in our experiments. The idea was that we could efficiently select pictures of people whose faces varied in attractiveness because hundreds of people would have rated them. After a few minutes of looking at these Web pages it was clear that this strategy was not going to work for women's faces in our study. Without conducting statistical analyses it was obvious that women's pictures showing breasts and cleavages had a big impact on men's ratings. These photographs simply could not be used in a study investigating facial beauty, because men were so distracted by breasts. Men prefer breasts that are firm and upward tilting, regardless of the size they prefer. This is the shape of breasts in young women who have not given birth, but they are also physical indicators of fertility.

Culture certainly affects how men react to women's bodies. However, cultural effects can interact with universal factors. In some cultures, men like heavier women and in others they like slender women. The very extremes are not liked in any culture. These cultural preferences are linked to the availability of food and other resources. In almost all developed countries that have reliable and rich sources of food, lower weight is associated with higher social and economic status in women. Fat countries like thin women. The relationship is the opposite in poorer countries where food is scarce. This phenomenon is called the "environmental security hypothesis." The general idea is that if food is scarce, a women's body fat indicates whether she has the energy reserve to bear children. Support for the environmental security hypothesis on attractiveness shows up in some striking examples. The physical characteristics of the Playboy Playmate of the Year from 1960 to 2000 track U.S. economic indicators. When economic times are difficult, the Playmates are older, heavier, and taller; they have larger waists, smaller eyes, larger waist-to-hip ratios, smaller bust-to-waist ratios, and larger body mass indices [36]. Similarly, between 1932 and 1995, American movie actresses with more mature features—small eyes, thin cheeks, and larger chins—were popular when times were tough, and those with baby-like features—large eyes, round cheeks, and small chins—were popular when times were plentiful [37]. When the going gets tough, the large get going in the eyes of amorous men.

Regardless of overall weight preference and the status given to women because of social and cultural conditions, one factor remains constant. Men prefer women with an hour-glass shape. This female shape with a narrow waist and large breasts and hips first develops around puberty. That means men prefer women with bodies that advertise their fertility. Men have waist-to-hip ratios between 0.85 and 0.95. Most fertile women have waist-to-hip ratios between 0.67 and 0.80. In fact, women with a waist-to-hip ratio under 0.80 are twice as likely to have babies as women with ratios greater than 0.80. The late psychologist Devandra Singh [38] found that men in many different cultures prefer women's bodies with waist-to-hip ratios around 0.70. Top female models' ratios often hover around 0.70. This preference for the ratio is true regardless of whether the culture admires slender or robust women. In the United States, both Audrey Hepburn and Marilyn Monroe are icons of beauty, despite being quite different in size. Both had waist-to-hip ratios of 0.70.

Given that we find both faces and bodies attractive, do we value one more than the other? Generally, people orient to faces more than bodies. However, men do vary in whether they choose to look at a woman's face or her body depending on whether they want a quick fling or whether they want to settle down into a serious relationship. In one laboratory study [39], young men were shown images of women in which the face and the body were initially covered. In choosing a partner, they were allowed to look at either the face or the body, but not both. For the fling, they looked at the body more often than the face, but for the serious relationship they looked at the face more often than the body. The insight motivating this study is that a woman's reproductive potential and fertility do not always go together. For example, a pregnant woman, barring mishap, is pretty likely to have a baby. But she is not fertile while she is pregnant. Women with low waist-to-hip ratios are fertile. So bodies are better signals for fertility than faces, and fertility drives men's desires for short-term partners. Men's bodies do not convey this kind of information, and in the study women did not vary in looking at men's faces or bodies depending on whether they were looking for short- or long-term partners.

Bodies move. Back in 1872, Darwin observed that we use dynamic cues to guess what other people are doing. How people move gives us a lot of useful information. Neurologists are trained specifically to observe the way people walk, because a person's gait gives us the best quick index of the nervous system's health. In the brain, as we shall see in the next chapter, we have areas that specialize in perceiving people's movement. We can recognize body movements without any information about their shape or color or contour. If you film a person walking in the dark with points of light fixed to 10 or 12 body joints and create what scientists call "point light walkers," people immediately recognize the points of light as moving human bodies. From just these moving points we can tell a person's gender, age, and whether they are anxious or relaxed and happy or sad.

Darwin thought that dance was a courtship ritual that signaled the quality of the dancer. Birds and insects dance. Female fruit flies choose their mates based on how well they dance [40]. Many male spiders attract female spiders with elaborate dances. Among funnel spiders, males that sway their abdomens fastest are most successful at attracting females [41]. One does not need to be a genius on a cruise ship to a distant land observing exotic animals or alien insects to appreciate that dance is a courtship ritual. One can see these mating rituals in any local nightclub. Even before getting to the dance floor, women, when interested in a man, move more often, more slowly, and with smaller amplitudes. Men are drawn to these come-hither movements.

Movement exaggerates some body parameters that are attractive when viewed statically. Movement can really display the efficient use of a symmetrical body. Middle-distance runners perform better than their asymmetrical competitors [42]. The hourglass shape of women's hip–waist ratio is emphasized by the alternating left–right sway of their walk. Women find point light walkers of symmetrical men attractive. Women rate very masculine point light walkers as most attractive when they are looking for short-term partners. This preference for masculine moving points of light is exaggerated when the women are ovulating [43]. In fact, ovulating women are also more likely to say "yes" when asked to dance by a man.

Bizarrely, it turns out that men's fingers are related to how women rate the attractiveness of dancing men. The ratio of the lengths of the ring to the index finger is affected by prenatal testosterone exposure. More testosterone exposure produces a longer ring finger in relation to the index finger. Men with larger ring-to-index finger length ratios are stronger and better at skiing, playing soccer, and sprinting. Apparently, they also dance better than men with lower ratios. In one study, video clips of male dancers with larger or lower ring-to-index finger ratios were shown to women.

The men that women thought were more attractive, dominant, and masculine based on these video clips had longer ring-to-index finger ratios [19].

So, it turns out that the parameters that make bodies beautiful are similar to those that make faces beautiful. We prefer bodies, like faces, that are symmetrical. We also experience bodies, like faces, that exaggerate sexual dimorphic features as beautiful. Men orient to signs of fertility in women. Women orient to signs of masculinity in men, which, as we shall see later, may signal the quality of the genes they carry. We don't know if averaged bodies are beautiful. Some cultural phenomena such as bodybuilding competitions, clothing styles, and dance often exaggerate the same parameters that underlie what we find beautiful. Finally, the context in which bodies are seen matters. Whether we are looking for a short-term or a long-term partner, and whether or not a woman is ovulating makes a difference in which bodies are thought to be beautiful.

The next two chapters will start our explorations of the brain. First, we will see how the brain works in general. This will seems like a detour from our journey into beauty, but it is necessary to establish some brain basics. Later in the book, I will integrate neuroscience information as we go along.

Chapter 5

How the Brain Works

The brain is an amazing organ. As a machine, it operates on about 25 watts of power, and yet it can do all the incredible things we do. There is no thought or fantasy or idea that does not play out in the brain. The brain has a hundred billion nerve cells with a hundred trillion connections. It is easily the most complex organ in the body. How can we possibly understand something so complex? The reality is that we have much to learn about the brain. But since the late nineteenth century, we have been accumulating knowledge about the brain from patients with brain damage, from electrical recordings of brain cells, and, more recently, from new ways of taking pictures of the brain.

The brain works with a logic that is tied to its anatomy. Understanding something about its anatomy and the way different parts of the brain are connected gives us insight into its operations. For our purposes, the structure and function of the brain open a window into what happens during aesthetic encounters.

Let's start with some basic brain terminology. The surface of the brain is called the cortex. It has grooves that are called sulci and ridges that are called gyri. The major parts of the cortex are called lobes. We have occipital, parietal, temporal, and frontal lobes. Major fissures separate different parts of the brain. The interhemispheric fissure separates the left and right hemisphere, and the Sylvian fissure separates the temporal lobe from the parietal lobe. Deep within the brain, clumps of nerve cells make up "subcortical" structures. The basal ganglia are one such clump that will be important to our discussion. The cerebellum is a separate and phylogenetically old part of the brain that lies below the occipital cortex.

There are two important principles to keep in mind when thinking about the brain. The first principle is that the brain has a *modular* organization. This means that different pieces of the brain specialize in carrying out specific operations. You can think of this organization like a car assembly line where each group of workers is trained to do specific tasks before the pieces are passed on to be further assembled, or "processed," by other workers. The second important principle is that the brain processes information in a *parallel and distributed* manner. Here, the assembly line analogy breaks down, because distant parts of the brain factory work together in a coordinated fashion. This principle means that the different areas that make up modules of the brain act together as part of a network choreographed to create most of our thoughts, feelings, and experiences. So, in order to understand how this complex organ gives us that fuzzy good feeling when looking at sunsets while taking long walks on the beach, we need to know something of its modular organization and its parallel and distributed processing.

At the most basic level, the brain has input systems, output systems, and things that modify whatever we take into our brain before we put something out. Information from the world comes into our brain from our different senses. Each of these senses, for what we see, hear, touch, taste, and smell, delivers information to different parts of the brain. Even though our eyes are in the front of our head, visual information goes to the back of our brain, into the occipital lobes. Different parts of the back of the brain are tuned to different parts of our visual world, such as color, shape, and contrast. These parts of vision are then combined into more complex objects, such as faces and bodies and landscapes, each with its own special area in the brain. These specialized areas are examples of the modular organization of the brain. One of the most striking clinical syndromes in neurology is called prosopagnosia. In this disorder, which happens because of the brain's modular organization, people have damage to the face area. They can read books, recognize objects, and navigate their environment. However, they cannot recognize faces, even those of their family and close friends.

Emotions have a big influence on how we process the information coming in through our senses. We all have the experience of being in a good mood and noticing sunny skies and chirping birds, or being in a bad mood and noticing dark clouds and pigeon crap all over the place. Our emotions color what we notice and how we experience them. Emotions are housed deep in the brain below its surface. These regions are called limbic areas. The limbic brain is responsible for our joys and fears, our happiness and sadness, our delights and disgusts. It is closely linked with the autonomic nervous system. This part of the brain is "autonomic" because it does its work tirelessly behind the scenes without our even being aware that it is humming along. The autonomic nervous system controls our heart rate and blood pressure and sweating responses, and links our brain and body in emotional experiences. This is why our pupils dilate when we are excited, our palms sweat when we are nervous, and our blood pressure rises when we are angry.

Meaning is another important system that profoundly affects how we see and experience the world. This point is obvious if we think about looking at the script of a language we don't know. For example, I can look at Arabic calligraphy and appreciate its graphic beauty without knowing what it means. However, if I could read Arabic, and the text were the story of Scheherazade from *A Thousand and One Nights,* my experience of these visual forms changes entirely. While this reading example is particularly dramatic, something along these lines happens whenever we look at most objects. Bringing knowledge to bears on whatever we are looking at has a huge impact on our experience of seeing. Meaning is mostly organized in the sides of the brain, in the temporal lobes. This is where general knowledge, our store of facts about the world, is stored. Besides general knowledge, we also know personal facts that refer to our individual histories. For example, knowing that the story of Scheherazade is a classic love story is different from knowing that I first heard the story in school as a boy in India. Personal memories are organized by a different part of the temporal lobe, tucked in deep, close to parts of the brain that control our emotions.

Finally, there are two big segments of the brain called the frontal and the parietal lobes. These structures have become larger in the human brain than in the brains of our closest primate relatives. They often work together, with the parietal lobes being important in what we choose to pay attention to and the frontal lobes being important in organizing our executive functions. These functions are so named because the frontal lobes are like executives of a big company. They tell other parts of the brain what to do and make plans that the rest of the brain might not be aware of.

In what follows, I will repeat all of this information, only in more detail. The detailed version includes long and complicated names of parts of the brain. Remembering the names of different brain areas has always been a chore even for medical and neuroscience graduate students. However, it would be silly to have "brain" in the title of a book and not talk about the brain with some specificity. Many of the neuroanatomical names will come up again when I talk about neuroscience experiments.

Visual processing starts in the retina of our eyes, where different kinds of nerve cells specialize to do different things. Cells called rods process luminance and cells called cones process color. When we colloquially say that beauty is in the eye of the beholder, we really mean it is in the brain of the beholder. So, we start in the brain at the occipital lobes. Visual information is sorted in different regions of the occipital and then the adjacent temporal lobes. For example, the shape of things, their movement or color, are all processed in different regions. This dismantling of our visual world into pieces gives us a Humpty Dumpty problem. How do we put all these pieces together again to give us our seamless visual experience of the world? Unlike all the king's horses and all the king's men, our brains manage to do just that. Exactly how is a topic that preoccupies many neuroscientists, but clearly it involves some kind of parallel processing. For now, let's note that different parts of our visual brain work on different things. There is an area that processes faces (the fusiform face area, FFA) and a separate area that processes places (the parahippocampal place area, PPA), including both natural and human-made environments. Close by, an area located on the side of the occipital lobes processes objects (the lateral occipital complex, LOC) in general. Next to it is an area that responds to the form of human bodies (the extrastriate body area, EBA). In this general vicinity of the brain a different area specializes in visual motion (area MT/MST, from middle temporal and medial superior temporal area). Still off to the side of area MT/MST and higher up is an area that processes moving bodies or biological motion (superior temporal sulcus, STS). So, we have a visual cortex that has specialized modules to process places, faces, bodies, and different objects. Is it a coincidence that much of visual art is about landscapes, portraits, nudes, and still lifes? Is it also a coincidence that we have an area specialized for biological motion and that dance is such a popular form of art?

As I mentioned before, limbic areas that process our emotions are sequestered deep within the brain. These areas do not lend themselves to being named easily as a "face" or a "place" area. Some of the main structures to be aware of as we think about aesthetic encounters are the following. The amygdala is an important part of the brain that handles emotions like fear and anxiety. It plays a role in coloring our memories with emotions, like remembering the anxiety you might have felt going into the Principal's office. The subcortical cluster of neurons that make up the basal ganglia have two big functions. One function is to work with the cerebellum and motor cortex to help coordinate movements. This function of the basal ganglia is impaired in patients with disorders like Parkinson's disease, in which people move stiffly and slowly, or Huntington's disease, in which people cannot control their movements. The second function of the basal ganglia is more germane to our discussion. The basal ganglia contribute to our experiences of pleasure and rewards. Important parts of the basal ganglia are the ventral striatum and one of its major subcomponents, the nucleus accumbens. These structures are washed in pleasure chemical signals, such as dopamine and opioid and cannabinoid neurotransmitters. The "high" that people experience from cocaine, heroin, and marijuana is a result of flooding these neurotransmitter receptors.

In the underbelly of the front of the brain lies the orbitofrontal cortex. This area is referred to as "orbito" because it is just above our eyeballs inside our skulls. This cortical structure is also tied to our experience of rewards. Other relevant parts of the brain for our discussion are the insula and the anterior cingulate. The insula harbors connections to the hypothalamus and together these structures regulate hormones and the autonomic nervous system. The anterior cingulate does many different things, like mediate pain and try to sort out conflicts we face. I will discuss these structures in greater detail when we ponder pleasure.

Meaning is often linked to language, which is organized in the left hemisphere of most people's brain. The area around the Sylvian fissure harbors language. Carl Wernicke, a famous German neurologist, in 1874 first reported that patients with damage to the back end of the Sylvian fissure, where the temporal lobe meets the parietal lobe, were unable to understand anything said to them. This location is now named

"Wernicke's area," to honor his discovery. People with damage to this area don't understand words.

Parts of the temporal lobes are critical storehouses of meaning. It is as if all bits of information about the world from our different senses—what we see, hear, and feel—got funneled into the sides of the temporal lobe and gathered together into our knowledge about the world. In a degenerative neurological disorder called semantic dementia, neurons in the left temporal lobe die, for reasons we do not understand. These patients gradually lose their knowledge of objects.

A small area tucked inside the temporal lobes, called the hippocampus, is critical to meaning that is tagged in time. Perhaps the single most famous case in all of neurology is a man named Henry Gustav Molaison, referred to as "HM" in medical writings. In the 1950s, HM had both hippocampi surgically removed as a treatment for his epilepsy. After this surgery he couldn't remember a thing, but otherwise was clearly very intelligent. Observations in HM led the way to our understanding of how general and personal meanings are organized in the brain.

I already mentioned that the parietal and the frontal lobes are larger in the human brain than in our closest primate relatives. The parietal cortex is well known for organizing the ways we think about space. It is like a spotlight that shines its beam of attention over different parts of our external world and helps guide us to reach for things and move through space. The frontal lobes encompass a huge amount of the brain. It organizes the information from the rest of the brain and prepares us to act in the world. The frontal lobes, along with our emotional centers, are where our sense of personality comes from. When people are neurotic, or extraverted, or laid back, these differences are written into differences in the frontal lobes and their connections with limbic areas. Generally, the frontal lobes are divided into three broad regions: the dorsolateral (on the sides), the medial (in the middle), and the ventral (the under belly) portions. The dorsolateral prefrontal cortex is where executive functions are housed. It is involved in making decisions and planning what we need to do. The medial frontal cortex more directly coordinates our motor systems and is involved in our sense of self. Damage to this area can cause a dramatic clinical syndrome called akinetic mutism, in which individuals appear awake but are completely unresponsive to the outside world. The ventral part of the frontal cortex allows us to regulate our own behavior, and part of this area is linked to reward systems. The orbitofrontal cortex is most important here. As we shall see later, parts of the orbitofrontal cortex that are closer to the midline of the brain are important for rewards, and parts further out to the sides for when we are sated with whatever was giving us pleasure.

To foreshadow the next chapter, this is what happens when we look at aesthetically pleasing objects. Information comes in from our eyes to the occipital lobes. This information is processed in different parts of the occipital lobe, which interact with our emotions in the limbic areas. When we like what we see, the pleasure or reward centers of our limbic areas are turned on. When we think about the meaning of what we are looking at, the temporal lobes are engaged. When we draw on our personal memories and experiences in aesthetic encounters, the inside of the temporal lobe comes online. As beautiful things engage us and capture our attention and we respond to them, we activate our parietal and frontal lobes.

Chapter 6

Brains Behind Beauty

I was having dinner with my friends Marcos Nadal, Oshin Vartanian, and Oshin's partner, Alexandra O. We were in Palma, Spain, at Marcos' invitation to talk about neuroaesthetics. Oshin is a cognitive neuroscientist who studies the brain bases of reasoning, decision-making, and creativity. He works for the Canadian Defense Department and is also editor of the journal *Empirical Studies of the Arts*. We were talking about science fiction and somehow got to talking about the *Alien* movies. I mentioned that I found Sigourney Weaver especially attractive. Oshin is a few years younger than I am. He prefers Wynona Ryder to Ms. Weaver. As he talked, his eyes glazed over, while Alexandra's eyes rolled back. She pointed out that Ms. Ryder was a shoplifter. For Oshin that observation didn't matter. After all, he insisted, "She's Wynona Ryder!" She didn't really mean to do "those things." Here was a neuroscientist, someone who works for a national defense department, who is an expert on human reasoning, unwilling to entertain the possibility that Wynona Ryder was culpable for her less than admirable acts. Of course, Oshin was being tongue-in-cheek with his insistence, but only in part. It turns out that Oshin is far from alone in resisting the idea that attractive people might not also be good people.

Truth, Goodness, and Beauty were three ultimate values for Plato. These values easily get mixed up, with beauty being associated with being good and true. Like Oshin, most of us think attractive people harbor all sorts of personal characteristics that, when you actually think about it, make no sense whatsoever. Attractive children are considered more intelligent, honest, and pleasant, and are thought to be natural leaders. In one study [44], teachers were given report cards of fifth-grade students, which included grades, work attitudes, and attendance, along with pictures of the students. The teachers expected the good-looking children to be more intelligent, sociable, and popular. Teachers often give more attractive children better grades, unless the tests are standardized. Attractive adults are thought to be more competent and have greater leadership qualities than less attractive people. They are expected to be strong and sensitive and better as politicians, professors, and counselors. They get jobs more easily and earn more money. Oshin would not be surprised to find out that attractive people are less likely to be reported for shoplifting even if they are clearly seen in the act. If caught, they are given lesser punishments. People are more willing to cooperate with or to help attractive people. These tendencies have been shown in planned experiments. People were more likely to return money found in a phone booth (back when such oddities still existed) to an attractive woman than to an unattractive one [45]. In another study, college applications were left in an airport with a note implying that the applicants' fathers was supposed to have mailed the application, but that the application had been dropped inadvertently. The applications, which were otherwise identical, had pictures of the applicants on them. People were more likely to mail in the application if the person pictured was attractive [46].

We usually don't realize that a person's attractiveness has a halo effect that makes us think well of them. Could it be that our brains sense attractiveness even when we are not conscious of doing so? Geoffrey Aguirre, Sabrina Smith, Amy Thomas, and I conducted a study using functional magnetic resonance imaging (fMRI) to answer this question [47]. fMRI technology enables us to see where in the brain blood flow is changing when a person is in a particular mental state. The changes in blood flow are a response to changes in underlying neural activity. Scientists design experiments to see which parts of the brain are active when people are engaged in different tasks. We had people look at pairs of faces to observe their brain's responses to facial attractiveness. The faces were generated by a computer program and were made so that they varied in how similar they were to each other and in how attractive they were. In one session people judged the identity of the faces, and in another session they judged their attractiveness. Designing the experiment this way allowed us to find out how their brain responded to attractive faces even when they were not thinking about beauty.

What did we find? When people were thinking about beauty, specific parts of the brain responded to more attractive faces. These areas included the face area (FFA) and the adjacent lateral occipital cortex (LOC) that processes objects in general. A skeptic might ask, if the entire visual brain becomes active, what have we learned? Well, in our experiment, not all parts of the visual cortex responded to facial beauty. The area tuned to places (PPA) didn't change as faces became more attractive, suggesting that the activity we saw was not a general response in visual cortex but one restricted to specific parts of the brain. Other researchers have also found similarly increased neural activity in these visual areas for more attractive faces. In addition to these visual areas, we found more activity in the parietal, medial, and lateral frontal regions of the brain when people judged beauty. We think these areas were engaged because people had to pay attention to the faces and make decisions about which faces they thought were attractive. For technical reasons, our scanning procedure was not sensitive to detecting neural activity in brain areas that are important for rewards. However, other investigators have found that these areas also respond to attractive faces. Attractive faces drive neural activity in the orbitofrontal cortex and part of the basal ganglia called the nucleus accumbens [48]. The amygdala has a complicated reaction to facial beauty. It reacts both to faces that are attractive and to those that are unattractive [49]. Later, when discussing our reward systems, we shall explore why the amygdala responds to the extremes of our likes and dislikes.

What happened when people looked at these faces and were not thinking about beauty? When they were thinking about the identity of the faces, the FFA and LOC (but not the PPA) in visual cortex continued to respond more vigorously to more attractive than to less attractive faces. Despite the irrelevance of beauty to what they were doing, their visual brains continued to react with changes in blood flow to these areas. Our visual brain reacts to facial beauty automatically. Kim and his colleagues were also interested in the question of whether the brain responded automatically to beauty. Their experiment focused on responses within the reward circuitry of the brain. They had people judge which of two faces were more attractive and which were rounder. Their strategy, like ours, was to see if brains react to attractive faces even when the people were concentrating on something else, like shape. They found that parts of the reward circuitry, specifically the nucleus accumbens and the orbitofrontal cortex, continued to respond to attractive faces even when subjects were pondering the faces' roundness [50]. Taken together, these studies point out that our brains are naturally tuned to facial beauty. In fact, our brains might always be responding to beauty around us. Perhaps we get little jolts of pleasure from beauty even when we are concentrating on other things. While it is difficult to always surround ourselves with beautiful people, these results make me wonder if we would be happier if we surrounded ourselves with beautiful objects.

As we saw in Chapter 4, people react to the beauty of bodies and the elegance of movement. What do we know about the brain's response to beautiful bodies? Unfortunately, we know very little. Despite the psychological studies of what people find attractive in bodies (reviewed in Chapter 4), I do not know of any neuroscience studies that examine our brains when we look at beautiful bodies. Building on what we know about faces, I predict that the more attractive a body is, the more activity there would be in the extrastriate body (EBA) and adjoining areas. Thus, when a person is consciously thinking of the body's beauty, parietal and frontal areas and the cingulate cortex would be active. Beautiful bodies would also make neurons in emotion and reward areas, such as the amygdala, nucleus accumbens, and orbitofrontal cortex, fire. The question of whether people would react implicitly to beautiful bodies they way they do to faces is difficult to anticipate. Perhaps the EBA and parts of the reward systems would still react to beautiful bodies, but the parietal and frontal lobes would not.

Static beautiful bodies are all over magazines, posters, and comic books. However, for the most part we see bodies in motion. How do our brains respond to beautiful movements, like dance? Brown, Martinez, and Parsons used positron emission tomography (PET) scanning to assess what happens in the brain during dance [51]. In their experiment, amateur dancers performed small cyclical tango movement while they were being scanned. The investigators examined specific components of dance: how people entrain their movements, how they follow rhythm, and how they move in predetermined spatial patterns. They found that a part of the cerebellum, called the vermis, fired when people entrained their movements to music. The cerebellum is an ancient brain structure that helps us maintain our balance. They found greater activity in the putamen for metric movements. The putamen is part of the basal ganglia that is involved in controlling movement. They found that parts of the parietal lobe were especially active when people moved their legs in a spatial pattern.

This study illustrates a general distinction that is worth keeping in mind when we consider aesthetic experiments. This distinction is between classification and evaluation. Studies in neuroaesthetics designed to focus on classification or evaluation answer different kinds of questions. You can classify something as an aesthetic object and then study its properties. That is what these investigators did. They took dance that we agree is an aesthetic object and studied the brain's response to its different components. By contrast, in an evaluative study, you could study the brain's response to movements (of any kind) as a person decides whether they like or dislike them. The object being evaluated isn't as relevant as our emotional reaction to it.

Beatriz Calvo-Merino and her colleagues conducted an evaluative study of movement. In this study, people watched 24 short dance movements. Half of the movements were from classic ballet and half were from Capoeira, the Brazilian dance-martial arts form. The people judged whether they liked or disliked the movements. People tended to like dance movements that involved jumps and whole body movements rather than smaller in-place movements of single limbs. The researchers found more neural activity in the right premotor cortex and in parts of the medial occipital cortex for movements that people liked than for those they did not like [52]. They suggest that these areas are part of the brain that organizes sensations and implements the movements that underlie dance. Here, it would seem that areas that implement dance are also involved in their evaluation. The pattern is analogous to our face study, where the area that classifies faces also responds more vigorously to the attractiveness of faces. Despite the fact that the way in which we classify objects is logically distinct from how we evaluate them, our brain might not make such a clear distinction in its operations. The same brain areas that classify objects get involved in evaluating them.

We are making headway in understanding the neuroaesthetics of faces and bodies. A general contour of how we might think about the brain and its response to beauty is emerging. The brain sorts different pieces of the world into different modules that carry out specialized processing. Some of these modules classify objects like faces and bodies and body movements. It looks like these same modules also evaluate these objects and probably work in concert with the brain's reward systems to produce our emotional responses regardless of whether they are delight or disgust. Many details of this entire system need to be worked out, but we are well on our way. In later chapters we will return to these systems in different contexts. For now, let's turn to the question of why we find objects beautiful to begin with. Why would our brain be tuned to beauty? Why are symmetry or averageness or sexual dimorphic features attractive in people? To begin to answer these questions, we turn next to Darwin's theories of evolution.

Chapter 7

Evolving Beauty

Shortly before starting to write this book, I vacationed in the deserts of Utah. The spare landscapes lay bare the vastness of time. In these deserts I found a useful analogy for evolution. Bryce Canyon is a small but very beautiful national park. Hordes of hoodoos populate the valley. Hoodoos are tall spires of rock that are usually capped by a hard stone. They occur in dry basins, formed by environmental erosion over many years. Hoodoos have variable thicknesses depending on the underlying stone and mineral layers. When seen in great clusters, as in Bryce Canyon, they look like crowds of looming figures, silent and solemn. Some Native American tribes would not enter these valleys for fear that hoodoos were hostile souls trapped in stone.

Hoodoos helped me envision how evolution works. Looking at them, I could imagine them actively emerging from the ground. The strongest spires pushed their way through the surface and reached for the sky. It is easy to think of "survival of the fittest" this way. But evolution does not work like that. Darwin's profound insight was that evolution selects traits passively over time. With hoodoos, harder stones are revealed as environmental vicissitudes remove softer strata. The shapes of the hoodoo spires result from erosion. Nature selects resistant stones and minerals for survival. No active agent devised a master plan to sculpt hoodoos. Similarly, nature selects resilient human characteristics, both physical and mental ones, by passively eliminating traits that are less effective than others for survival and reproduction. Over many generations, physical and mental traits that give people even a small advantage in producing healthy children accumulate in larger proportions of the population.

Darwin also realized that natural selection couldn't explain many oddities he observed in animals. The stag's antlers, the antelope's horns, the peacock's tail, the flamboyant colors of birds and fish challenged the theory of natural selection. These extravagances encumber the animal and attract predators. They could not possibly enhance survival. But the same extravagances also attract the attention of potential mates. Darwin realized that another force, sexual selection, was also at work. The idea of sexual selection is that attractive animals are more likely to mate and produce more offspring. Adult male and female creatures adapt their appearance to get more or better partners. Courtships in the animal world involve some (usually males) competing for their partners and others (usually females) being choosy about the partner they will mate with.

As a historical aside, Darwin's views of sexual selection took much longer to be accepted within the scientific mainstream than his ideas of natural selection. He pointed out that males in most species compete with each other and advertise themselves to females. Females then chose worthy males. Part of the resistance to the acceptance of sexual selection was the cultural climate of the Victorian age. The idea that sex was fundamental to our evolution and that women were driving players in this unfolding drama was even harder to accept than the idea that there was no God to create us in His image.

But back to the hoodoo analogy and its relation to sexual selection. Let's imagine that hoodoos are indeed animate. They mate and produce new little hoodoos. A hard capstone is regarded as attractive by some hoodoos. So, hoodoos with hard-looking capstones get to mate more often and pass on this trait. As a result, later generations have more and more hard-capstone hoodoos. Hard capstones are a "fitness indicator," which means that they indicate the hoodoo's fitness to resist the environmental beating, even though amorous hoodoos might not be aware of the survival implications of a hard head. The larger and more extravagant capstones might be more attractive, but they come at a cost. The capstone could get so heavy that the spire breaks and collapses into a heap. The point short of collapse where characteristics seem outrageous and are costly to maintain is where things get really interesting. The evolutionary psychologist Geoffrey Miller [53] has argued that sexual selection more than natural selection explains many human "indulgences," like arts and culture.

Thus, two forces, natural selection, which enhances survival, and sexual selection, which enhances reproduction, offer some insight into why we think some objects are beautiful and others are not. When it comes to people, we evolved to seek mates that maximize the survival of our children. I should be clear, however, that what drives most people to sex is desire and pleasure, rather than a cold calculation about genes catapulted into the indefinite future.

Imagine an early era of amorous hoodoos. Some found hoodoos made of soft, curvy limestone attractive. Some found very narrow-waisted hoodoos attractive, and others found hard-headed hoodoos attractive. Of those with these three physical features, the hard-headed hoodoos were most likely to survive. The soft, curvy ones eroded, and the narrow-waisted ones collapsed. A preference for soft curves and narrow waists ended up not producing more little hoodoos. So, the preference for hard heads and the expression of hard heads were more likely to be inherited by the next generation. The hoodoos did not have to learn that hard heads indicated fitness. In each subsequent generation this proportion increased, until having a hard head became a "universally" attractive feature. Carried over to people, the attributes that indicate fitness and those we find attractive survived and grew proportionately. This preference is an outcome of the fact that people find pleasure in and desire for people based on the way they look; those looks happen to correlate with their fitness.

As I mentioned before, men rank physical beauty as very important in what they want in a partner. Unbeknownst to them, the features that men find beautiful in women are linked to fertility and the likelihood of producing healthy children. Women are also interested in physical beauty, but they rate other features of men more highly. These differences in how men and women rank their preferences for the opposite gender have been found in almost every culture that has been investigated [20]. As the saying goes, "Handsome is as handsome does." Women are choosy about their partners. They care about a man's attractiveness, but also about their social position, prestige, and wealth.

As we have seen, three parameters contribute to physical attractiveness in people. These are averageness, symmetry, and the exaggeration of sexually dimorphic features (those that distinguish men and women). How do these three parameters relate to evolution?

Averaged features are the central tendencies in a population. By definition, they are not extremes. As in hoodoos, extreme physical features are typically not very healthy. As a result, averageness ends up being a general sign of health and fitness. A mate with averaged features would have a greater chance of producing children that are likely to survive. We know this implicitly, given the general impression that in-bred people "look funny" and people of mixed race look attractive. One explanation for this impression is that physical mixtures of diverse populations are a sign of greater genetic diversity. Greater genetic diversity implies greater flexibility to adapt and survive under more varied conditions. Are people with averaged features healthier than others? We don't really know, because most people these days are pretty healthy through their reproductive years. An important point to which we shall return later is that adapted traits gave an advantage to our distant ancestors dealing with very different environmental conditions from those we face now. "Fitness" today is not relevant to whatever fitness meant for our ancestors.

Another explanation for the effects of averaged features rests in the way our brains process information. The evolutionary mechanism here is natural rather than sexual selection. With natural selection, certain abilities give slight advantages in survival. One such ability is the quick formation of categories when we look at many different examples of any particular thing. So, it is helpful to know that a dachshund, a terrier, a husky, and a Labrador are all dogs. A strategy in forming a category quickly is to establish a prototype [54]. Prototypes are examples of what we think are typical of a category. For example, a robin is a prototype for a bird in the way that an ostrich is not. Prototypes are often an average of many examples of the category. People prefer prototypes of different kinds. This preference has been shown when people are asked to rate colors and musical pieces [55]. Prototypes, because they are typical of a category, are processed more easily and liked. This general property of the mind, liking prototypes, also applies to faces. Averaged faces are prototypes of a population.

Symmetry is the second parameter that we find attractive in faces (and bodies). The evolutionary argument for symmetry being attractive follows a logic similar to the argument for averaged features. Symmetry is also a fitness indicator. Symmetry reflects a healthy nervous system, since many developmental abnormalities produce physical asymmetries. Symmetry also indicates a healthy immune system. Parasites, which played an important role in human evolution, produce asymmetries in most plants and animals and humans. Humans differ in their susceptibility to parasites, based on the genetically determined strength of their immune systems. Thus, facial and bodily symmetry advertises a person who is resistant to parasites. While attractiveness is highly regarded in every culture, Gangestad and Buss found that attractiveness is valued even more highly in cultures with serious infestations of malaria, schistosomiasis, and other virulent parasites [56].

Asymmetrical bodies, besides indicating infections or developmental anomalies, are also less efficient in physically moving toward desirable goals and avoiding dangers. Earlier, we saw that symmetrical middle-distance runners perform better than asymmetrical runners [42]. Skeletal remains from prehistoric Native Americans show that older individuals had more symmetrical bones than younger individuals [57]. This observation is striking, because aging introduces more asymmetries. It probably means that symmetrical people were healthier and survived longer than their asymmetrical cousins.

Sexual dimorphism is the third parameter for finding certain physical features attractive. These features also give their bearers advantages in sexual selection. Feminine features in women signal greater fertility. Men are attracted to women with facial features that advertise their fertility, which end up being features that combine youth and maturity. Faces that are too baby-like might mean that the woman is not yet fertile or might not be able to raise her children. Women need some degree of sexual maturity to bear and raise children. Thus, men find women with big eyes, full lips, narrow chins (indicators of youth), and high cheekbones (indicator of sexual maturity) attractive [58]. The idea is that younger women have a longer time to have children than older women. Men who are drawn to young, fertile women are likely to have more children than men drawn to older women. This preference is likely to be passed on and to accumulate in subsequent generations.

The physical features that make male faces attractive also have evolutionary explanations. Testosterone produces masculinized features. In many different species, testosterone suppresses the immune system. So, the idea that masculine features would be a fitness indicator makes no sense. Here, the logic is turned on its head. Rather than the fitness indicator, or "good gene," hypothesis for masculine features, scientists invoke a "costly signal" hypothesis [59]. Only men with very strong immune systems can pay the price of testosterone on their immune system. The most commonly cited example of a costly signal in the animal world is the peacock and its tail. Certainly this cumbersome but beautiful tail doesn't exactly help the peacock to move with agility, to approach desirable peahens quickly, or to avoid predators efficiently. Why would such a handicapping appendage evolve? The basic reason is that the peacock is advertising to the peahen that it is so strong that it can afford to spend all this energy maintaining a costly tail. Consistent with this idea, the most colorful birds are found in areas with the most parasites [60], again suggesting that especially fit birds are showing off by diverting resources to these extravagant appendages. The logic of these costly signals explains what to me has been a profound puzzle. That puzzle is why some men will spend thousands of dollars on something like a Rolex watch when a simple Timex, and now cell phones, do the job. Spending a lot of money on expensive watches, cars, and homes has nothing to do with their utility. They are costly signals displayed by men to indicate their fitness, in the hopes that choosy women will like their pretty tails.

With masculine features for which the immune system takes a hit, the argument is that the owner of this face is so tough that he can get away with spending some fitness capital on his chiseled jaw. He has fantastic genes to contribute. So women should find these testosterone-infused masculinized faces more attractive. And they do. In fact, both women and men who are concerned with germs are even more likely to find masculine features attractive than if they are not thinking about susceptibility to infections [61]. Somewhere deep in the heterosexual ancestral brain, the link between a strong immune system and masculine features was forged.

As we saw before, most women find very masculine faces attractive only to a point. Faces advertise more than fitness. Hyper-masculine faces also advertise dominance. Women are invested in their children and also want partners who will share in that investment. Hyper-masculine faces advertise that their owners might not be cooperative partners and good parents. So, women end up liking men with masculine features that have been somewhat feminized (as can be done in the laboratory [62]), because this combination signals that the man has good genes and will offer long-term support and be good parents to their children. Overall, for women, this combination is a better package than a hyper-masculine or an extremely feminine-looking man.

With the ovulatory shift hypothesis [63], we saw that women shift their preferences according to their needs. This phenomenon was shown dramatically in an isolated hunter-gatherer group, the Hadza, living in Northern Tanzania [64]. The investigators manipulated the recordings of the pitch of male and female voices, knowing that higher levels of testosterone are associated with lower pitched voices. Men preferred women with higher pitched voices as marriage partners. Both men and women viewed lower pitched voices as coming from better hunters. Women who were breastfeeding showed a preference for higher pitched male voices, whereas women who were not breastfeeding preferred lower pitched male voices. Women who were not nurturing infants prefer men that can go out and score the big kill. Once women have an infant, they prefer a man who will be invested in raising their child. Even though this example is not about a reaction to faces or bodies, I mention it because it makes the general point that women's preferences vary depending on the context in which they engage men. Testosterone-infused physical features like big Dick Tracy jaws or low Barry White voices are more attractive as short-term partners and not necessarily as men with whom to settle down.

The hypothesis that women's choices and desires have the subtext of wanting children with the best genes, along with the most resources available, has fascinating consequences, as we saw earlier with the ovulatory shift observations. When women are most likely to get pregnant, they desire more masculine men, presumably to invite stronger immune systems into their genetic mix. Also, if young women are asked about whom they find attractive for long-term or short-term partners, their choices shift in the direction of desiring more masculine-looking men as short-term partners. For long-term partners, women want their men to be manly with a little added warmth and commitment. Underlying these observations may be the idea that women are selected to be pretty polyandrous and choose different partners at different times for different reasons.

Curiously, the size of men's testicles supports the idea that women are choosy and moderately polyandrous [65]. Primates have varied social structures. Gorillas typically have a dominant male who guards over a harem of females. Male gorillas compete with each other to control the harem, but only one of them wins. Female gorillas are mostly restricted to mate with this strong and fit male. Chimpanzees sort themselves differently. A female chimp might mate with as many as 50 males when she is ovulating. In this case, male competition is happening at the level of the sperm. The sperm that wins the race for the egg wins. One way to increase the chances of your sperm winning is to produce a lot of them and hope that the numbers work in your favor. To do that, you need big testicles. Male gorillas are selected to be big and strong with massive muscles, but they have tiny testicles. Their sperm stroll into the picture only after the competition is over. Male chimpanzees, by contrast, are not nearly as big, but they have huge testicles. Their competition begins after ejaculation. The relative size of human male testicles is somewhere in between gorillas and chimpanzees. Women, by implication, are neither as monogamous as female gorillas, restricted to one big bull of a man, nor are they as polyandrous as female chimps, having sex with as many as 50 males during the peak of ovulation.

Monkey love aside, the evolutionary logic for beauty is that attractive features have survived because they are relatively good indicators of health. If this is true, then many of these indicators should correlate with each other. To test this idea, Grammer and colleagues [58] listed 32 attractive features (such as lip, eye, breast size, body mass index, waist-to-hip ratio, body color, skin texture, averageness, symmetry) in 96 American women. They found four components that contribute to what men find attractive. The first two that correlated negatively with attractiveness were body mass index and babyness-androgeny. The latter two that correlated positively with attractiveness were what they called nubileness and symmetry/skin color. Grammar and colleagues suggest that decisions are easier when these factors point in a similar direction. It doesn't matter which of the cues one happens to notice; the strength of the cue, rather than its content, becomes relevant. In their mathematical models, the lowest correlated feature increased the degree to which attractiveness was predicted. They speculate that, rather than approaching attractiveness, what we are really doing is avoiding features we find unattractive. Generally, it is true that most people are more inclined to avoid risks than to seek rewards in many life decisions. So this hypothesis certainly has merit. At the same time, it is hard to imagine that Romeo and Juliet were drawn to each other simply because they found all the other Montagues and Capulets unappealing.

How does culture influence these parameters of human beauty? To understand at least one way that culture comes into play, we turn to herring gulls. Many years ago, the ethologist Tinbergen [66] observed that herring gull chicks get adult gulls to regurgitate food by tapping on a red spot on an adult gull's yellow beak. The chicks also peck at a red dot on a yellow stick if that stick replaces a beak. If more red dots are painted on the stick, the chicks peck even more vigorously even though they have never actually seen such a strange object in nature. This phenomenon of an exaggerated response (the peak response) to exaggerated versions (the shifted version) of a stimulus that would evoke a normal response is called the "peak shift."

Many cultural practices involve peak shift responses to sexually dimorphic features. The neurologist Vilayanur Ramachandran has pointed out that Hindu temple sculpture makes use of this peak shift principle [67]. These images depict goddesses with big breasts and waist-to-hip ratios as low as 0.3, he thinks, to exaggerate the power of their fertility. Exaggerations of sexual dimorphic features are the stock in trade of comic books. Supermen have big square jaws, huge muscles, and ultra-V-shaped torsos. Superwomen have big eyes, big breasts, narrow waists, and wide hips. We refer to the highest paid models as supermodels. Measurements of parts of their face are those of girls under 10 years of age. Talk about a peak shift in expressing youthfulness!

In classical ballet, we have also seen a peak shift principle gradually emerge over the last 60 years. Body postures that were considered fixed and static actually shifted over the years. Body positions are increasingly vertical and leg angles increasingly extreme. These recent forms that are gross exaggerations of the original classical postures are preferred by naïve ballet viewers [68].

Fashions and cosmetics often work on this principle of taking features that we evolved to find attractive and then turbocharging them to produce a peak shift response in the observer. According to Nancy Etcoff, author of the book *The Survival of the Prettiest*, we spend twice as much money on personal-care products and services as on reading material. These personal-care products usually enhance the size of eyes or make the lips fuller or emphasize higher cheekbones. In other word, they exaggerate features that men find attractive in women.

The obsession with enhancing the attractiveness of our features has been with us for a very long time [69]. In Southern Africa, archeologists have found red ochre sticks that are many tens of thousands of years old. They think these sticks were used to adorn bodies [70]. Ancient Egyptians had well-developed cosmetics practices. In King Tutankhamen's tomb, 3,000-year-old skin moisturizers were found. The ancient Egyptians stored moisturizers and had formulae to prevent wrinkles and blemishes. Most men and women shaved their bodies, and shaving sets from 2000 BCE have been found. They used red ochre to paint their cheeks and lips and henna for their nails. In the Indus Valley, cosmetics were used as far back as 2500 BCE. Different skin masks were used for different seasons, hair removal products were common, lip color and methods of dental hygiene were used widely, and products to prevent premature graying of hair were available. In Ancient Greece, precious oils, perfumes, cosmetic powders, eye shadows, skin glosses, paints, beauty unguents, and hair dyes were used commonly. Ancient Rome inherited beauty practices from the Egyptians and Greeks. Women used cosmetics imported from other parts of the world. Eye makeup and rouge was common. Various juices, seeds, plants, and other materials were used to make the skin appear whiter. Women would even bathe in ass's milk, which served as a chemical peel.

The ancient practices of enhancing our physical attractiveness have their contemporary expression in the cosmetic surgery industry. In 2010, over 13 million cosmetic surgical procedures were performed in the United States. And it is not just young white Hollywood starlets that want these procedures. Men are one of the fastest growing segments of the population wanting physical enhancement procedures. African-American, Hispanic, and Asian Americans are also in on the act. There is no mass market for procedures to make us look mature or wise or kind or sincere or witty. The procedures are designed to make us look beautiful. In fact, most cosmetic procedures correct asymmetries or emphasize sexual dimorphic traits.

To summarize our foray into evolution and beauty, we evolved to find certain features about people beautiful [71]. Our ancestors who happened to find pleasure in these features were also the ones that were more likely to pass on their genes into the future. We inherited their pleasures and their sense of beauty. These very same features can be exaggerated by evolution as costly signals and peak shifts. Culture can certainly modify what we regard as beautiful, but it does so especially successfully when it exaggerates sexual dimorphic features. By using the peak shift principle, these cultural modifications exaggerate our responses to things that are etched in our brain to be regarded as beautiful.

Next, we turn to places rather than people. How do we explain our fondness for places? Why are some places more attractive than others? Do any of the principles that apply to beauty in people also apply to beauty in places?

Chapter 8

Beautiful Landscapes

John Muir, the intrepid naturalist, said, "Everybody needs beauty as well as bread, places to play in and pray in, where nature may heal and give strength to body and soul." The idea sounds dramatic, yet, I share the sentiment that wilderness provides a deep comfort, especially during hectic and difficult times. I am not alone in this view. Across many studies, scientists find that people prefer natural to built scenes [72]. When feeling stressed, a walk in the woods helps calm most people in a way that a stroll through built environments simply does not.

Eighteenth-century aesthetic theorists focused on nature. Landscapes could be beautiful, or sublime, or picturesque. The questions they asked are similar to those that we still wonder about. Why do we get excited or awestruck or calmed by places? Could the beauty of places have anything to do with the beauty of people? To state the obvious, places are not people. It is hard to see how parameters like averageness, which play a critical role in faces, apply to environments. Symmetry could be relevant to formal gardens and built environments, but hardly to natural landscapes. Some spots are romantic, but it stretches the imagination to think that sexual selection drives our preference for places in the way that it drives our desire for people. What makes a place beautiful?

Strong evolutionary forces selected minds that find some places more beautiful than others. Powerful emotional responses evolved to guide and encourage actions that improved our ancestors' chances of surviving and reproducing. Maybe the places that we find beautiful are the very ones that improved our ancestors' chances of survival. These preferences would have evolved in the Pleistocene era, the long swath of time from 1.8 million to about 10,000 years ago. Our Pleistocene hunter-gatherer ancestors moved frequently. They covered different terrains and had to decide where to go, where to stay, and when to move. Anthropologists think these groups preferred environments that could be explored easily and that provided resources needed to survive. One such environment in Africa is the savanna. The savanna has areas of slight elevation that give long, unimpeded views. Trees are relatively scattered, allowing one to easily see large roaming mammals from a distance. Large mammals provide much-needed protein. The trees themselves offer protection from the sun and could be climbed to avoid predators.

People like pictures of the savanna, even if they have never been there. In one study [73], people of different ages (8, 11, 15, 18, 35, and over 70 years) looked at images of tropical forests, deciduous forests, coniferous forests, deserts, and the East African savanna. Eight-year-old children said that they would like to live in or visit the savanna more than any other environment. After the age of 15, people also liked deciduous and coniferous forests. Since none of these people had visited savannas, the implication is that this preference was programmed into their brains. This programmed preference is called "the savanna hypothesis." As people get older, this preference gets modified by where they have lived. Synek and Grammer confirmed these findings in a different study [74]. They showed that young children in Austria prefer landscapes sparsely dotted with trees and with low mountains, scenes that resemble the savanna. After puberty their preferences shifted to areas with denser trees and higher mountains, again, presumably, because of more experience with these terrains.

Trees themselves support the savanna hypothesis. Japanese gardeners have developed a sophisticated aesthetic of tree forms. They select and prune their specimens to create specific shapes that coincidently mimic the characteristics of trees in the savanna. One study looked at what people like about a tree that grows in the savanna [75]. *Acacia tortilis* is a medium-to-large tree that is also called the umbrella thorn because it has a distinctive wide crown. The tree has different forms depending on the richness of the environment in which it grows. Different forms of the tree were shown to people from the United States, Argentina, and Australia. All three groups preferred trees with moderately dense canopies and branches that split close to the ground. This is precisely the kind of *Acacia tortilis* that grows in areas with the most resources. To borrow a term we used to describe features of attractive people, these trees are fitness indicators. They indicate how fit an environment is in providing for human needs.

The savanna hypothesis is romantic. It invites us to imagine that we humans are yearning for home, expressing a collective unconscious desire to return to our ancestral roots. Having spent some time in Botswana, I share a romantic sense of "Africa." The landscape is spare. You get the sense that it has not changed for hundreds of thousand years. Despite its untamed quality, the land draws you in. The experience of the land is the opposite of something exotic. Curiously, the pleasure comes not from its novelty but from its deep familiarity. The savanna hypothesis tells us that certain landscapes are widely regarded as attractive. However, the savanna hypothesis cannot be the whole evolutionary story for landscape preferences. Humans moved out of Africa to populate virtually every continent. If our ancestors couldn't have survived in places other than the African savanna, migration would not have gotten very far. In the vastness of Pleistocene time, people must have evolved preferences for other landscapes.

What makes a landscape attractive? We like places that provided safety and sustenance to our hunter-gatherer ancestors. These places have water, large trees, a focal point, changes in elevation, relatively open spaces, distant views of the horizon, and some complexity [76]. These features are characteristic of savannas, but they are also found in other environments. Orians and Heerwagen point out [77] that fundamental questions for our hunter-gatherer ancestors were where to go and then whether to stay or to continue exploring. After choosing to enter and stay in an area, our ancestors had to gather local information. They had to be vigilant for signs of predators and be on the lookout for sources of water and food. Kaplan and Kaplan found that people now like scenes with a combination of features that predict safety and nourishment in an environment [78]. Scenes need to be "coherent" so that relevant information can be gathered quickly. For example, repeated forms and relatively uniform areas give a scene coherence. Without coherence, a scene is difficult to read and dangers difficult to anticipate. Simultaneously, scenes need to be somewhat complex. Complexity is the richness and number of elements in a scene. Without complexity, a scene is boring and unlikely to promise much in the way of food and water. Moderately complex scenes also have a quality that the Kaplans called "mystery." Mystery tantalizes us with the possibility of interesting discoveries if we only dared to explore. Roads or steams that glide around hills and take the viewer around partially obstructed views, enticing us to enter and find out whatever lies around the bend, give a scene this sense of mystery.

The time of a scene, in addition to its spatial layout, has a big impact on its beauty. At dawn, in the Grand Tetons, if you go to the Moulton Barn (probably the most photographed barn in the world), you will see hordes of photographers setting up their tripods. The same scene looks more or less beautiful at different times. The beautiful times are those that required the attention of our itinerant ancestors. The shifting light at dawn or dusk is salient for safety in a world of nighttime predators. Salient changes also include sudden shifts in weather. For example, certain cloud patterns or quick changes in lighting might be a call to move. These patterns are found in scenes we regard as beautiful. Time also changes an environment at a slower pace. Seasonal changes require responses that involve anticipation and planning. Budding trees and first greens are clues to near-future abundance. Flowers are particularly interesting natural features. Very few flowers are eaten and yet they are highly prized as beautiful objects. Flowers signal that an area will have good foraging in the near future. Flowers, like the shapes of trees, are a landscape fitness indicator. Much research still needs to be done on landscape preferences. But as a general principle, spatial and temporal features that were signals for safety and sustenance to our ancestors are what we now regard as beautiful.

What do we know about the neuroscience of landscapes? As I mentioned earlier, a specific part of the visual cortex, the parahippocampal place area (PPA), is tuned to respond to natural and built environments more than to faces or bodies or other individual objects. This area works with another area called the retrosplenial cortex (RSC) to organize the space in which we move around. My colleague, the cognitive neuroscientist Russell Epstein, showed that the PPA represents local scenes that are viewed directly. These scenes could be landscapes, cityscapes, room interiors, and even "scenes" made out of Lego blocks. Neurons in the place area are not affected by whether a place is familiar. Viewing or imagining a scene also activates the RSC. However, unlike the PPA, the RSC reacts more vigorously to familiar than to unfamiliar places [79]. This observation suggests that the RSC helps recover the memory of a scene. The RSC is also interconnected with other important areas of the brain that organize space, such as the posterior parietal cortex. Epstein suggests that the RSC integrates where we are with our remembrances of that location to give us a rich sense of place.

The passionate environmentalist Edward Abbey said, "A journey into the wilderness is the freest, cheapest, most non-privileged of pleasures. Anyone with two legs and the price of a pair of army surplus combat boots may enter." How does our brain respond to the pleasure of places? To investigate the neural underpinnings of landscape preferences, neuroscientists Yue, Vessel, and Biederman [80] had people look at different scenes while lying in an fMRI scanner. These scenes included natural vistas, city streets, and rooms. They found that the right PPA responded more vigorously for scenes that people said they liked than for those that they did not like. Similarly, they found more neural activity in the right ventral striatum. Again, as with beautiful faces, we find a pattern of greater activity for beautiful places in an area of visual cortex that specializes in processing places in general. These observations suggest that this region classifies as well as evaluates scenes. It does this evaluation by coordinating its neural activity with areas that encode pleasure and rewards.

Stepping back from all these studies, what have we learned so far? Five principles underlie our sense of beauty in people and places. First, similar to faces and bodies, our preferences for places are partly hard-wired. We prefer vistas that resemble savannas even if we have never visited such a place. These preferences are then modified by later personal experiences. Second, our Pleistocene ancestors who were drawn to places that also happened to improve their chances of survival passed on these tastes in what we now regard as beautiful. Natural selection rather than sexual selection played the dominant role in the evolution of place preferences. Third, the brain's responses to beautiful landscapes involve neuronal ensembles in the visual cortex that classify environments, and these areas fire together with neurons in reward systems. It is too early to be sure, but the evidence suggests that our visual brain not only classifies things, it also evaluates them. Fourth, we respond to fitness indicators. In faces, these could be big eyes, full lips, or square jaws. In landscapes these are trees that indicate a bountiful environment or flowers that promise rich sources of nutrition. Fifth is the role of enhancements. We saw earlier that cosmetics have played a long role in human history. Generally, cosmetics, including invasive plastic surgery, enhance physical features that we evolved to find attractive. Is there anything analogous to human environmental creations? Gardens are examples of landscape enhancements. They are designed to delight and give pleasure. They often exaggerate the aspects of natural landscapes that we find beautiful, by leaving open places, multiple vantage points, partially hidden paths, and flowers that signal the promise of bounty.

Despite the differences between people and places, we see that common principles account for their beauty. Next, let's push such comparisons to an extreme, by looking at numbers and math. We don't think of numbers as evoking sensations, and yet people think that certain combinations of numbers are beautiful. How could that be? Does beauty of numbers have anything in common with beauty in people and places?

Chapter 9

Numbering Beauty

In my lab, we study the neural bases of human cognition. We conduct fMRI experiments with young, healthy people and behavioral experiments with people who have had brain injury. A typical study might enroll between 12 and 20 people. Recently, we completed a study with 17 participants. The experiment worked. I liked the design of the study, the data were informative, and the conclusions were interesting. Yet, I was not happy. The reason for my unhappiness was the number 17. It did not feel like the right number for an experiment. This feeling had nothing to do with what statisticians call "power," which is determining the number of people needed in an experiment to be confident about the validity of its results. My discomfort was specifically about the number 17. Sixteen seemed like a good number; so did 20. Eighteen seemed pretty good, but 19 gave me the same discomfort as 17. For reasons I cannot articulate, numbers that are divisible in several ways feel more right for an experiment than prime numbers.

I have no idea if other people share my sense of "rightness" of numbers for experiments. The fact remains that I have preferences for numbers in specific contexts, as does everyone who confesses to having lucky and unlucky numbers. The Pythagoreans ascribed all sorts of social attributes to numbers. The number 1 was the generator of all things; 2 was a feminine number, and 3, a masculine one. The number 4 was the number for justice and order. Five represented the union of the first female number 2 and the first male number 3, and so it was the number for love and marriage. Attributing social characteristics and values to numbers is taken to great lengths in the many systems of numerology. Numbers are not desiccated abstract entities confined to realms of pure thought. We can like and dislike them. We can also find them beautiful.

Why bother with math in a discussion of beauty? The main reason to consider math is to find out why something so different from people or places can be beautiful. We don't have sex with numbers or want to live (at least concretely) in them. Numbers and formulae are not sensuous and seem far removed from the eighteenth-century view of aesthetics as experiences grounded in sensations. Yet, people talk about math in the same way that they talk about other beautiful objects. Consider these comments by Bertrand Russell, the British mathematician and philosopher, from *The Study of Mathematics*:

> Mathematics, rightly viewed, possesses not only truth, but supreme beauty—a beauty cold and austere, like that of sculpture, without appeal to any part of our weaker nature, without the gorgeous trappings of painting or music, yet sublimely pure, and capable of a stern perfection such as only the greatest art can show. The true spirit of delight, the exaltation, the sense of being more than Man, which is the touchstone of the highest excellence, is to be found in mathematics as surely as poetry.

I will illustrate ways in which math might be beautiful and how this relates to beauty in people and places. Even though I am no mathematician, intuitions about beauty in numbers make sense to me. At a first pass, math can be beautiful in two ways. First is the way that numbers pervade nature. Certain mathematical relations keep popping up in the physical and biological world. We often experience the revelation of this underlying mathematical structure in nature as beautiful. Second is the way that numbers behave. Numbers interact, come together and fall apart, and lead to surprising conclusions that we also experience as beautiful. Along the way, we will also discuss whether these mathematical properties are out there waiting to be discovered or whether they are creations of the human mind.

A number regarded as very beautiful by many mathematicians is the never-ending, never-repeating number 1.6180339887.... Isn't its beauty obvious? The number is better known as phi, or the golden ratio (Figure 9.1). It was discovered by a Greek mathematician, Hippasus, in the fifth century BCE, and later elaborated on by Euclid. The golden ratio is shown by a line divided into two segments in which the ratio of the whole line to the longer segment is equal to the ratio of the longer

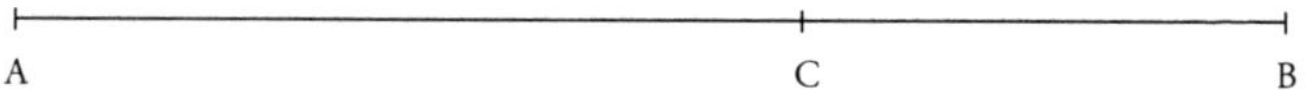

A C B

Figure 9.1. The golden ratio. The ratio of the whole line AB to the longer segment AC is equal to the ratio of the longer AC to the shorter segment CB.

to the shorter segment. Phi is an irrational number, meaning that it cannot be described by a ratio of two whole numbers. Legend has it that the discovery of irrational numbers caused great consternation among the Pythagoreans, for whom numbers were a central feature of the rational organization of the world.

Phi has captured the imaginations of mathematicians and historians, perhaps like no other number. Claims about phi abound, as described vividly by Mario Livio in his book *The Golden Ratio*. The number may have been used in the design of constructions in Mesopotamia, burial sites in Egypt, including most famously the pyramids, as well as in the proportions used in the Parthenon in Athens. The name of the number phi comes from Phideas, the architect who designed the Parthenon. The golden ratio is thought to give these classical structures their harmonious beauty.

Phi has seemingly magical properties. If you square 1.6180339887...you get 2.6180339887.... If you take its reciprocal you get 0.6180339887..., that is, these numbers have the same set of nonrepeating endless numbers after the decimal point. Phi is related to a group of numbers called the Fibonacci series. Fibonacci was a mathematician in Pisa who published a book called *Liber Abaci* in 1202. In this book, he posed the following problem. A pair of rabbits is placed in an enclosure. They produce a pair of offspring every month. The offspring become productive in their second month and produce offspring at the same rate. At the end of the year, how many rabbits are in the enclosure? The solution is depicted graphically as follows. Each *R* represents a mature pair of rabbits, and each *r* a young pair. The sequence of rabbits would be

January	R
February	R r
March	R r R
April	R r R R r
May	R r R R r R r R
June	R r R R r R r R R r R R r

From the series you can see that the number of adult pairs each successive month are 1, 1, 2, 3, 5, 8... The number of young pairs follows the same sequence but is offset by one row, 0, 1, 1, 2, 3, 5... and the total number of rabbit pairs follows the same series also offset by one row in the opposite direction, 1, 2, 3, 5, 8, 13... Every number in the series is the sum of the preceding two numbers. To answer the question of how many rabbits would be present in a year, one simply doubles (because these are pairs of rabbits) the twelfth number of the series.

The great astronomer Johannes Kepler discovered a fascinating relationship between the Fibonacci series and the golden ratio. If you express each successive number in the series as a ratio, you get

1/1 = 1.00000
2/1 = 2.00000
3/2 = 1.500000
5/3 = 1.666666
8/5 = 1.600000
13/8 = 1.625000

and so on, with numbers that get closer and closer approximations of phi. For example, further down in the series you have 987/610 = 1.618033.

Phi and the Fibonacci series show up in nature in surprising ways [81]. For example, the spiral arrangement of leaves on a stem of plants like the hazel, blackberry, and beech is arranged at every one-third turn. For the apple, apricot, and live oak, leaves are arranged every two-fifths of a turn, for the pear and weeping willow, every three-eighths of a turn. These fractions are made exclusively of numbers in the Fibonacci series. Pineapples show intriguing arrangements that draw on the Fibonacci series as well. The surface of pineapples is made of hexagonal scales. Each hexagonal scale is part of three spirals of different steepnesses. Most pineapples have 5, 8, 13, or 21 spirals, all numbers in the Fibonacci series.

New leaves on a stem often emerge at about 137.5 degrees of angle. This number forms a golden ratio. That is, 360 degrees, or the complete turn, can be divided into two sections, 222.5 and 137.5 degrees, that give (222.5/137.5 = 1.64) the golden ratio. These angles are referred to as the golden angle. If you closely pack points separated by 137.5 angles on a tightly wound spiral, the eyes notice two families of spirals, one moving clockwise and the other counterclockwise. The numbers of spirals in each direction are usually consecutive numbers in the Fibonacci series. This phenomenon is demonstrated beautifully in the florets of a sunflower: most have 34 and 55 (both numbers in the Fibonacci series) spirals going in opposite directions. To continue with the plant theme, the numbers of petals in a daisy tend to be Fibonacci numbers, such as 13, 21, or 34, and the petals of a rose overlap with each other in a way that forms multiples of phi.

The shape of nautilus shells, the horns of rams, and the tusks of elephants all form a related famous spiral described by the mathematician Jacques Bernoulli. Even hurricanes, whirlpools, and giant galaxies have this spiral. In this spiral, the radius grows logarithmically as the spiral moves around its curvature. The logarithmic spiral is related to the golden ratio in the following way. If you take a golden rectangle and portion off a square within it, you are left with a smaller rectangle that also has the golden ratio. If you snip off another square within this, you also get another smaller golden rectangle. This is called a self-similar pattern, because the geometric relationships are identical at different scales. If you connect the points at which these successively smaller squares divide the golden rectangle, you get a logarithmic spiral (Figure 9.2).

Why should plants, shells, and even hurricanes care about these related numbers and spirals? Leaves placed along a spiral that follows the golden angle are packed most efficiently. At this angle, buds can never line up exactly on top of the other and the space around the stem is used maximally. Another clue to why the golden angle appears in nature comes from physics. Douady and Coulter [82] dropped magnetic fluid into a dish of silicon gel in a magnetic field that was stronger at the periphery than in the center. The small magnets repelled each other and were pushed radially by the magnetic field gradient. When they settled into an equilibrium, they converged onto a pattern in which each of the drops was separated by the golden angle. Thus, these arrangements and spirals seem to minimize energy in systems. The buds appear at these points along the stem probably to minimize the energy requirements of buds that would naturally repel each other because of their needs for sun and nutrients from the stem. Similarly, the spirals probably grow or find themselves in equilibria that have the least energy demands to maintain their structure.

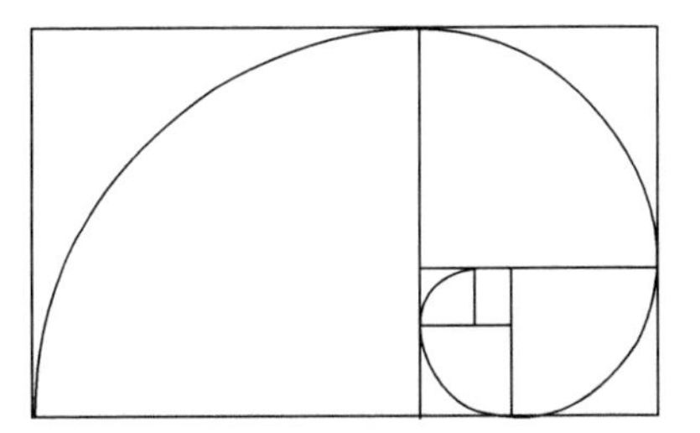

Figure 9.2. The golden (logarithmic) spiral. The golden rectangle is formed by two sides comprised of the golden ratio. Portioning off a square within the golden rectangle leaves a smaller golden rectangle, a pattern that can be repeated ad infinitum. Connecting the points of the successively smaller squares gives the golden spiral found in nautilus shells, rams' horns, whirlpools, and galaxies.

To summarize this little stroll into the wonderland of numbers, the ratio phi is irrational and beautiful, because it combines and splits into surprising and beautiful combinations, and it reveals hidden regularities in the world. Phi is related to the Fibonacci series, in that a ratio of each successive number in the series gets closer to the irrational phi. Phi is related to the golden angle, which is formed by the same ratio of angles in a circle as the ratio of segments of a line. Phi is related to the golden rectangle, which has the property of looking exactly the same at different scales and is related similarly to the logarithmic spiral. If these odd relationships were not bizarre enough, they show up in the leaves of plants, the petals of flowers, the shells of mollusks, the horns of rams, the patterns of hurricanes, and the shapes of galaxies!

These observations suggest that math is something objective out there, waiting to be discovered. Math is objective in two senses. First, math reveals truths about our physical world that are described by remarkably simple quantitative relationships. So, as Newton pointed out, force equals mass times acceleration ($f = ma$), and Einstein showed that energy was related to mass and the light constant ($e = mc^2$). These simple relationships were true about the world before humans evolved and would be true even if humans had not evolved. Second, math reveals relationships that are unconditionally true. $2 + 2 = 4$ is true in a way that is unambiguous and remains true in all contexts. That $2\pi r$ describes the circumference of a circle was true in the age of the dinosaurs as it was when *Homo erectus* walked the earth and as it will be when we own personal jet-packs. These relationships were waiting to be discovered and passed on by Babylonians, Egyptians, Greeks, Indians, Arabs, and Italians.

What are the properties of beautiful math? As evident in the number phi, beautiful math is revelatory. It is succinct, uses minimal assumptions, surprises us with novel insights, and generalizes to solve other problems. Euler's identity, $e^{i\pi} + 1 = 0$, is regarded by many mathematicians as the most beautiful theorem. Carl Friedrich Gauss, the famous mathematician and one of the founders of modern statistics, claimed that if Euler's identity was not apparent to students upon being told it, they were not likely to become first-class mathematicians. Sadly, I am not a first-class mathematician. Why is Euler's identity beautiful? Each basic arithmetic operation—addition, multiplication, and raising to a power—occurs once. It links five fundamental mathematical constants, the number 0 (additive constant), the number 1 (multiplicative constant), the number π (ubiquitous in Euclidean geometry and trigonometry), the number e (the base of natural logarithms used widely in scientific analysis), and the number i (an imaginary unit of complex numbers connected to algebra and calculus). Euler's identity is beautiful because it is succinct and surprisingly general.

Jürgen Shmidhuber reformulated the idea that beautiful math reveals hidden regularities, as the appreciation of data compression [83]. We experience pleasure when we recognize this compression. Things that are too regular are not beautiful because they are obvious. Things that are too complex and have no regularity are not beautiful because they are chaotic and overwhelming.

Another view of math, besides being something objective waiting to be discovered, is that math is a human creation. We do all sorts of things with numbers, like count sheep and build computers. Like other human creations, some math is beautiful. As we shall see, specific parts of the brain are specialized to process numbers and mathematical relationships. Numbers could be like language. We use letters symbolically and then have rules that govern how we put them together in words and sentences. Letters, for the most part, are purely symbolic, and how we use them tells us more about the structure of our mind than about the structure of the world. In the same way that we can combine words and sentences beautifully, maybe we can combine numbers beautifully.

What do we know about the neuroscience of math? People who have damage to the left posterior parietal lobe can have a condition called dyscalculia. They cannot do even simple arithmetic operations. In the 1920s, Josef Gerstmann, an Austrian neurologist, described patients with dyscalculia, along with a cluster of three other symptoms: finger agnosia (loss of knowledge of one's own fingers), confusion about the terms left and right, and dysgraphia (problems with writing) [84]. It is tempting to think that these symptoms co-occur because they constitute a meaningful unit. One could imagine that our fingers relate to counting since we use a base 10 counting system that almost certainly derives from the number of our fingers. We also conceive of a number line with spatial dimensions specifically along the left–right horizontal axis. Perhaps our sense of numbers is related to this left-to-right spatial layout. Finally, writing uses purely arbitrary symbols put together in ways that are themselves not arbitrary, similar to the ways in which numbers are put together. In fact, much of Europe used the Roman system of having letters signify numbers until Arab scholars brought the number system from India to the West. While it is difficult to prove in a way that would satisfy a mathematician, the physical topography of the brain is designed to maximize efficiencies. Having different operations located in such a way that they can share subroutines makes sense. While the symptoms that Gerstmann described are probably not causally related (e.g., finger agnosia does not produce dyscalculia), they probably share subroutines, which is why they are processed so close to each other in the brain.

In his book *The Number Sense,* the cognitive neuroscientist Stanislas Dehaene argues that numbers are a fundamental form of knowledge that allows us to parse the world. He argues that in a world of movable objects, evolutionary pressures would have encouraged the ability to efficiently quantify objects. Monkeys and preverbal human infants can approximate quantities. They can also do simple addition and subtraction, like 1 + 1 = 2, or 2 – 1 = 1. Animals and humans show a "size effect" when it comes to discriminating numbers. Larger pairs of numbers need greater differences to be discriminated easily. Discriminating between 5 and 7 is easier than between 35 and 37. Thus, animals and humans have fuzzy representations of quantity that allow quick approximations and comparisons. This representation of quantity is different from the precision of numbers, which is built upon our symbolic abilities. In fMRI studies [85], the intraparietal sulci of the parietal lobes in both hemispheres are active when people approximate quantities. The left inferior parietal cortex is active when people perform calculations regardless of the kind of numerical notation, suggesting that an abstract symbolic system is encoded in this area. This is also the area that, when damaged, produces dyscalculia.

You may have noticed in this discussion of the neural basis for math that we have not talked about aesthetics. I am not aware of any studies that have examined the neuroaesthetics of math. However, based on what we know of the neural basis for beauty in faces and landscapes, what might we expect? We would expect neural activity in regions of the brain that process numbers coordinated with neural activity in brain regions that process pleasure and rewards. When numbers are represented as symbols, we expect the left posterior parietal cortex to be active. When numbers are represented as approximate quantities, we expect the intraparietal sulci in both hemispheres to be active. When mathematical relationships are involved, we expect the parietal cortex to be active in concert with parts of dorsolateral prefrontal cortex that maintain and manipulate complex information. These areas that process numbers and math would fire along with reward areas. We should keep in mind Russell's description of math's "cold and austere" beauty. At one end of the pleasure continuum are the rewards we experience looking at a beautiful body or face, and at the other end are the rewards of understanding a beautiful theorem. The first activates our desires, the second activates a liking response. Liking is a reward that is not necessarily tethered to desire. This distinction between wanting and liking is important in thinking about aesthetics and is one we shall revisit in the section on pleasure.

Why should numbers give us any kind of pleasure? If we start with the premise that pleasures drive evolutionary adaptations, then the question is what about enjoyment of math gave humans a survival advantage? From evolutionary theory, the only two choices are the forces of sexual selection and those of natural selection. It seems unlikely that numbers directly affect sexual selection. Of course, as mentioned before, heterosexual men find women with waist-to-hip ratios around 0.70 very attractive. So this is a numerical parameter. But there is nothing about 0.70 that is intrinsically attractive. It simply represents ratios in a curve that signals fertility. The evolutionary pleasure in the beauty of numbers must be driven by natural selection.

Why would pleasure in numbers be adaptive? The reasons I offer are at best speculative. In the Pleistocene era, humans must have been able to quantify things and predict quantities of future things. For areas with wild game, knowing how much meat might be available relative to the number in a group would have been critical to deciding whether to stay or press on. Predicting areas of good foraging based on the growth of edible plants would have been an important survival skill. People who enjoyed quantities, probabilities, and correlations would have had an evolutionary advantage in meeting their needs to assess immediate and future sources of nourishment and shelter.

A more general evolutionary advantage for taking pleasure in math would be in seeing patterns in what would otherwise be an overwhelming amount of information. The ability to reduce information to succinct quantitative relationships with broad generalizable properties is a skill that would have enabled early humans to quickly glean important information from their environment. The discovery of underlying structural relationships in the world would have helped them master their surroundings. The simpler the final formulation, the easier and more useful it would have been in the mental toolkit of our ancestors. Those ancestors who found pleasure in playing with mathematical relationships, who enjoyed seeing underlying patterns in complex environments, who could capture these relationships succinctly, improved their chances of survival. We find pleasure in these desiccated mathematical objects because our ancestors that experienced such pleasures were the ones who survived and begat us.

Chapter 10

The Illogic of Beauty

I started my musings about beauty by pointing out its mysteries. Beauty is all around us. We are drawn to it and yet we don't understand it. Where is beauty? Is it in the world or in our heads? Is there just one type of beauty, or is it simply a trick of language that we call both a fashion model and a mathematical theorem beautiful? Is beauty universal, or is it culturally constructed? Is our experience of beauty a burning passion or a cool contemplation? Most mysteriously, why is there such a thing as beauty? Having taken a little tour of beauty in people, places, and proofs, perhaps we can answer these questions.

Beauty is not to be found exclusively in the world or in our heads. Our minds are part of the world, and how we think and experience and act has been molded by the world over eons of evolution. The experience of beauty comes from the interactions between our minds and the world. Our brains evolved to find some objects universally beautiful. By universal, I mean that humans share this sense of beauty. Most people from different cultures consider the same objects beautiful. Parameters that contribute to beauty in faces, like symmetry and averageness, are universally appealing. Even infants respond to these features as if they were beautiful. We have seen that most people prefer similar landscapes, such as savannas, even when they have never visited these places.

Even if experiences of beauty are shared universally, cultural influences affect these experiences. Cultural influences often build on universal biases by interacting with or enhancing their effects. The kinds of women's bodies that heterosexual men find attractive are an example of an interaction of universal and cultural influences on attractiveness. As a universal principle, men find women with waist-to-hip ratios of 0.70 particularly attractive. However, culture influences whether men prefer wider or narrower women. In poor countries and in difficult economic times, men prefer large women's bodies. In rich countries and in times of excess, men prefer thin women's bodies. However, regardless of whether men prefer wider or narrower women, they prefer a waist-to-hip ratio of 0.70.

Cultural influences on beauty often exaggerate universal biases in our sense of beauty. Here, the peak shift phenomenon is at work. Take something that evokes a response and then exaggerate it to enhance the response. This strategy drives the cosmetic industry. Makeup and various accessories exaggerate appealing features. Similarly, most cosmetic surgeries in men and women enhance sexually dimorphic features that we regard as attractive. Stylized approaches to bodies as in bodybuilding competitions, Hindu temple sculpture, Barbie dolls, and comic book superheroes all exaggerate sexual dimorphic features. Similarly, constructed landscapes like some gardens and golf courses often exaggerate environmental features to which we are drawn.

Finally, contextual effects influence what we regard as attractive. In talking about faces and bodies, I emphasized erotic beauty. But the context in which we see faces influences their attractiveness in the literal sense of capturing our attention. As we saw earlier, in some contexts, physical beauty may not be most salient in an experience, for example, if we are on the lookout for danger. Our responses to faces are influenced by the context in which we encounter them. As we shall see later, context has a profound influence on how we experience art.

When we respond to beauty in people, places, and even mathematical proofs, are we responding to the same property? In the brain, the experience of beauty arises from how we process the sensory properties of objects, the meanings we associate with those objects, and the objects' interactions with our emotions and reward systems. Our brains have a divide-and-conquer strategy when it comes to processing the world. The visual system processes different visual elements—like color, and shape, and luminance, and movement—in different parts of our brains. The elements are then concatenated into objects that also have specialized bits of their own real estate in the brain. So the visual cortex processes faces and places in different regions. Moving bodies and even numbers have their own locations in other parts of the brain. We activate these specialized parts of the brain when we look at these different objects aesthetically.

Faces and landscapes are very different from each other and are processed in different parts of the brain. Math doesn't even have obvious sensory qualities to be processed. Math is pretty far removed from the idea of aesthetic experience as embedded in sensations. Thus, sensory processing in beauty differs for different objects. At this level of processing, beauty cannot be the same for different objects.

Is there something common to the emotional response to beauty? In the next section, we shall look more closely at the pleasure we derive from beauty and the nature of our brain's reward systems. We probably have a range of rewards that for now I shall refer to as lying on a continuum from hot to cold. Hot would be pleasures that evoke our passions, such as desiring a sexually arousing body. Cold would be pleasures that evoke a more detached contemplation, like appreciating an elegant proof. So objects are beautiful and probably tickle our reward systems in different ways. Beauty is not a single property, but a collection of different properties of objects in the world that mix and match in flexible ways to give us the experience of beauty.

Now that we have deflated the idea of beauty as a single property, can evolutionary arguments give us a common thread to bind together all that is beautiful? First, let's review the basic evolutionary argument and take some pains to clarify one version of evolutionary arguments that by itself does not make much sense. Consider the following description of the evolution of beauty, by Arthur Krystal, from *Harper's Magazine* (September 10, 2010):

> ...basic aesthetic preferences that, as a matter of Darwinian adaptability, cause us to be attracted to certain shapes and sounds as opposed to others. In sum, the argument is as follows: Whatever helped the first humans survive most must have appealed to them, and this knowledge of what was beneficial was programmed into their brains and inherited by subsequent generations. Our aesthetic preferences, therefore, are the result of evolved perceptual and cognitive abilities, and though the pleasure associated with beauty is no longer essential for survival, it continues to influence how we feel about both art and nature.

Variations of such descriptions are standard fare when it comes to evolutionary explanations for beauty. However, this description fails to express a critical nuance. As stated, it says that early humans liked things that helped them survive. This knowledge then got programmed into their brains. Now, even if those things no longer help us survive, we still find them appealing and regard them as beautiful. The argument doesn't add up. I appreciate the usefulness of the Philadelphia subway system but I do not find it beautiful like the Washington, DC metro system. Why should something useful be regarded as beautiful? And how does that knowledge insert itself into the brain? The argument that something that is useful is also beautiful is too neat and logical. This description misses the critical point that there is *nothing logical* about what we find beautiful.

Evolutionary usefulness accompanies beautiful objects but does not cause their beauty. The hoodoo preference story I used earlier illustrates this point. In the mythical hoodoo past, some hoodoos found hard heads attractive, others found narrow waists attractive, and yet others found soft and rounded curves attractive. There is no inherently logical reason that one kind of hoodoo is more beautiful and gives more pleasure than another. However, hoodoos that liked hard heads had more progeny because hard heads survived more than narrow waists and soft curves. Thus, the preference for hard heads was passed on, and the proportion of hoodoos that liked hard heads grew while the proportion of hoodoos with other preferences dwindled. The "universal" preference did not start out universal. It ended that way. The preference for hard heads did not arise because it was useful or adaptive. Rather, the pleasure in a feature *that happened to be adaptive* is what survived. This pleasure became more common in the population because other pleasures, which were equally illogical to begin with, did not survive over time.

Evolutionary adaptive arguments do not give us a unified concept of beauty. If we look at what is regarded as beautiful in people, places, and proofs, we see different driving forces. Sexual selection drives much of our preferences for people. Natural selection drives much of our preferences for places. A different form of natural selection, the enjoyment our nerdy ancestors found in processing and compressing large amounts of information, drives our preference for numbers. We are left with a collection of reasons for why we find things beautiful, rather than a single reason that corrals all these experiences together.

Beauty is a mongrel. It is a collection of different properties that engage different parts of the brain. Beauty produces different responses and evolved within us for different reasons. Beauty engages our sensations, emotions, and meaning flexibly. While beautiful objects happen to be useful, it is pleasure that drives us to beauty. Now, we are ready to take a closer look at this crazy little thing called pleasure.

PART II

PLEASURE

Chapter 1

What Is This Thing Called Pleasure?

Cats are hedonists. We can learn much from them. If they understood language, they would regard the notion of "guilty pleasure" as patently absurd. My cats revel in their pleasures. Zizou could find that one patch of sunlight to wallow in as its warmth spread on her inky fur. Reno arches her back and scrunches her eyes in bliss as her belly is rubbed. Whorf relished the smell of food. He would sit on the counter across the stovetop, eyes hooded, sniffing the air as spices wafted off simmering sauces.

Like cats, our sources of pleasure are many. Food and sex are among the most basic. But so many other things give us pleasure: watching a sunset, winning a bet, taking a nap, accomplishing a goal, listening to music, dancing, laughing, learning. We can lose ourselves in pleasure. Pleasures seduce us into setting aside rational thought.

The sheer variety of objects that give us pleasure raises questions. These questions are similar to the ones we asked about beauty. Is there a common currency for all pleasures? Is there a pleasure center in the brain? Are we always aware of our pleasures? Should we enjoy pleasures with abandon, or should we be cautious and deny them? Why do pleasures sometimes go bad? Why do bad things sometimes give us pleasure? Why do we enjoy Mae West's quip, "When I'm good, I'm very good. When I'm bad, I'm better."

Our pleasures are rooted deeply in our evolutionary past. Our ancestors that found pleasure in objects with survival value were the ones of their era to produce more children. We usually approach objects that we find pleasurable. Even single-cell organisms like amoebae are driven to approach and avoid things. They react to their chemical environment by approaching things they need and avoiding things that are toxic. This is the most basic survival strategy for all mobile creatures. The approach-and-avoidance strategy evolved into reflexes as organisms became more complex. In mammals, approach and avoidance are linked to homeostasis, the processes of keeping our internal environment (like temperature and hydration) relatively constant in a changeable environment. Apart from these basic survival functions, approach-and-avoidance behaviors evolved to ensure survival of progeny.

Pleasures are an important kind of reward. Psychologists call rewarding objects "positive reinforcers." These objects encourage us to repeat behaviors that deliver rewards. Food, water, and sex are primary rewards. Things like money and art are secondary rewards. We learn to derive pleasure from secondary rewards.

The potency of pleasure as a driver of behavior in mammals is obvious in my cats. We humans, more so than cats, can temper our pleasure-addled brains. One way to keep ourselves in check is by framing our pleasures. We can change our emotional experience of objects by altering the context in which we consider them. This human propensity is well understood by religious institutions that admonish us to be wary of our "base" pleasures. Framing means that we are not slaves to our sensations. Our cognitive systems can reach down into our pleasure centers and rejigger our pleasure experiences, as we shall see later when we talk about brand labels and fetishes. Another way that humans temper pleasure is by considering the passage of time. If I place a bowl of food in front of my cats, they do not ruminate over the fact that I have been traveling a lot, so perhaps it would be wise to save some food for later. Hedonists that they are, they have little regard for delayed gratification. By contrast, any medical student, presumably among the smartest of our youth, is a grand master in delayed gratification. They are willing to forego years of adequate sleep, regular meals, and decent pay for some future reward.

In what follows, we shall examine different kinds of pleasures, with an eye to how they are relevant to beauty and art. To start with, we shall look at food. Food fulfills one of our most basic survival needs and is a source of great pleasure. The pleasure of food is embedded in the chemical sensations of taste and smell. Then, we will turn to the pleasure of sex.

Here sensations also reign, but the approach behavior is complicated, as recounted in endless romantic comedies and soap operas.

From these appetitive pleasures we shall move to pleasure from something that we neither eat nor with which we cuddle. I am talking about money. Money also drives our behavior profoundly despite being abstract. Money represents pleasure for what it gets us, even though it is just metal or paper or plastic. Money offers a window into the dynamics of delayed gratification between when we get money and when we spend it. The use of time as a wedge between us and our pleasures is what economists call "forward discounting." The idea is that we can choose to deny ourselves immediate rewards because we expect bigger rewards in the future.

Finally, pleasure is tied intimately to learning. Why does pleasure help us learn? Recent work in neuroscience is uncovering startling facts about how we learn that extend from associating juices with tones to judging the reputation of others.

Thus, our sources of pleasures are many. Pleasures are embedded in sensations and can be modified by cognition. We can lose ourselves in pleasures or step back from them. These experiences of pleasures are relevant to understanding our experiences of beauty. As we shall see later, they are also relevant to understanding our experiences of art, but in less straightforward ways. We shall see what pleasures mean in the brain, in behavior, and, ultimately, in our aesthetic experiences.

First things first: food.

Chapter 2

Food

"To eat good food is to be close to God." Primo makes this pronouncement in the movie *Big Night*. Primo, played by Tony Shalhoub, is an uncompromising immigrant Italian chef who has to contend with "philistines" who order a side of spaghetti with their seafood risotto. Primo does not know that good food, besides filling our mouths and bellies, sets off a cascade of events in the brain that tweak dopamine, opioid, and cannabinoid receptors, the major chemical conveyors of our pleasures. Primo does know that God is bliss.

The bliss of food is driven by our chemical senses, taste and smell. Gradients in the chemical environment were among the earliest signals that drove single-cell ciliated organisms to swim toward nourishment. Millions of years and billions of neurons later, we are still led by our noses. Our chemical senses reach into ancient parts of our brain. They go directly to areas that process emotions and pleasures. Even though we can detect and discriminate hundreds of smells, we are terrible at describing them. Unlike other senses, smell refuses to be tamed by language.

The pleasure of food combines the senses of taste and smell along with touch in the mouth. The five major tastes are sweet, salty, sour, bitter, and savory (or umami). These tastes evolved to help us recognize foods that nourish or threaten us. Sweet signals sources of energy. Salty helps us maintain our internal chemical environment. Sour helps us keep a proper acid–base balance. Bitter warns of the danger of toxins. Savory (the characteristic taste of monosodium glutamate) points us to protein. In a very real sense, taste and smell guide what we should or should not put into our bodies.

The basic predisposition of what we should eat seems pretty hardwired. Babies lick their lips to sweet tastes and gag at bitter ones. By the age of 3, children have quite universal preferences for certain smells [86]. For example, they like strawberry, spearmint, and wintergreen, and they dislike butyric acid (vomit/cheese) and pyridine (spoiled milk). Despite the hard-wiring of our tastes, our preferences for what we eat can be modified, even in fetuses. We have known for years that fetuses develop taste receptors in the third trimester. Before ultrasounds were available, X-ray contrast was injected into the amniotic fluid to assess the fetuses' health. If saccharine was injected with the contrast, fetuses increased their swallowing, and if a bitter substance was injected fetuses decreased their swallowing [87]. During the third trimester, the smells or tastes that the mother experiences cross the placenta and are experienced almost immediately by the fetus. Early exposures influence preferences that the baby will later express. Babies respond to the smell of alcohol more robustly if their mother drank alcohol frequently during pregnancy than if their mothers did not consume alcohol [88]. Children of human mothers that taste anise during the third trimester are more likely to like the taste of anise than if their mothers had no anise during this period [89]. More generally, fetuses exposed to different tastes are more likely to be open to new foods later in life.

Anyone who has had a bad cold knows that food is not simply about the taste on our tongues. The pleasure of food lies in its flavor. Flavor is a complicated perceptual experience that combines smell and taste. While our noses may do poorly compared to some of our pets, the neurologist Jay Gottfried points out that we routinely detect trace odorants in the range of parts per billion. We can distinguish odorants that differ by one carbon atom and discriminate tens of thousands of distinct smells. Odorants bind to receptors in our nose that send information to be integrated in the brain in a structure called the olfactory bulb. At the olfactory bulb, different smells generate different patterns of neural activity. This information is relayed onto a part of the cortex called the pyriform (meaning pear shaped), which begins to connect the outside world with our internal experiences. The pyriform cortex is tuned to the identity of chemicals anteriorly and the quality of odors posteriorly [90]. By identity, I mean the molecular and chemical composition of the odorant. The anterior pyriform cortex offers a snapshot of the outside chemical world that we smell. By quality, I mean the integration of information that produces the perception of smell. The posterior pyriform cortex reflects our subjective experience of these smells. This subjective interpretation can be modified by experience and the context in which we encounter smells.

Information from the olfactory bulb passes to other parts of our brain with little filtering. Vision, audition, and touch pass through a deep brain structure called the thalamus. The thalamus filters sensory information before it gets to the cortex. By contrast, smell signals bypass the thalamus on their way to the cortex and get to our pleasure centers pretty directly. The olfactory bulb sends its signals to other areas in the brain, such as the anterior olfactory nuclei, the olfactory tubercle, the amygdala, and the entorhinal cortex. The pyriform cortex sends information to the orbitofrontal cortex (OFC).

The deodorant industries know that we can counteract unpleasant smells with pleasant ones. Unpleasant smells activate the posterolateral OFC and pleasant smells, the medial OFC [91]. The amygdala plays another role. It responds to the intensity of the smell rather than whether the smell is pleasant or unpleasant [92]. The amygdala response seems to be designed to get us moving in either direction, to approach or to avoid whatever is producing smells in the environment. Together, these parts of the brain are engaged when we sense what we like and what we don't like and when we are motivated to move toward or away from those objects.

Smells arrive at the receptors of our noses in one of two ways. They get there directly when we sniff objects, or they get there indirectly through the back of the throat from smells given off by food in our mouth. We experience these two "smells" very differently [93]. Smells directly from the nose tell us about objects out there in the environment. They invite us to seek the sources of pleasant smells and move away from fetid ones. Taste, by contrast, tells us about something that is already inside our mouths. To survive, it is critical that we not misread the signals from taste, to make sure that we eat the right things and not poisonous ones.

The indirect pathway for smells is critical to the experience of flavor in our mouth. The smell from food coming from inside our mouth is integrated with taste and other sensations. Touch plays an important role in the mouth. The creaminess of ice cream, the crunchiness of potato chips, the heat of chilies, the astringency of wine are all part of the rich orchestration of our flavor experience.

Taste, like smell, gets to the brain pretty directly. Information from the tongue taste receptors is carried to various centers in the brainstem. These brainstem centers then send the information to the insula, the amygdala, the hypothalamus, and the hippocampus. Together these brain areas integrate taste with other senses and with the chemical environment of the body and, like smells, taste rewards are encoded in the OFC and the ventral striatum [94]. Taste is the most basic of rewards given that food and drink are fundamental to our survival. In scientific parlance, taste has "intrinsic value."

Have you ever been around someone who ate so much pizza that they complained about feeling sick? The complaint is not just about their belly feeling full; it is that the smell of pizza or the thought of putting another slice in their mouth (for the time being) feels disgusting. Such experiences of taste and smell show clearly that pleasure we get from the same food can change dramatically. Experiments that use satiety (the sense of fullness) to investigate pleasures of smell and taste reveal how this works in the brain. For example, in one experiment, people smelled banana or vanilla odors before and after eating bananas to satiety [95]. To start with, they liked both odors equally. In another experiment, people drank tomato juice or chocolate milk to satiety [96]. Again, to start with, they liked both liquids equally. In the experiments, after being sated, people had less neural activity in the medial OFC to these smells and tastes that they intrinsically liked. At the same time, they had more neural activity in the lateral OFC, an area normally activated when people feel aversion. So, a person who likes both chocolate milk and tomato juice, when sated to chocolate milk, will activate the lateral OFC to chocolate milk and the medial OFC to tomato juice. The satiety experiments demonstrate that the OFC is sensitive to the history of our pleasures. The fact that satiety is specific to tastes and smells is the reason that we still have room for dessert after we are completely satisfied with the savory tastes of a fantastic main course. Here, sweet is not sated, and the medial OFC is still open to sweet, sweet pleasure.

Pleasures help us learn. The pleasure of food is a classic way to make associations, as Pavlov showed in the 1890s. He trained dogs to associate various sounds like bells, whistles, and metronomes with food. These dogs learned to salivate to the sounds even before they received food in their mouths, and later they salivated even when food was not present. This kind of learning is called "classical conditioning." Smells are also effective in classical conditioning. Faces paired with pleasant smells are discriminated more quickly than faces paired with aversive smells [97]. Since knowing what to eat is critical to our survival, it is hardly surprising that other senses latch onto taste and smells when we need to learn. The conditioning effects of pairing other sensations to taste and smell activate the medial OFC. This means that when a sensation that is not intrinsically pleasurable becomes pleasurable by association, it activates the same area of the brain that is activated by the originally pleasurable sensation.

There is a special delight in rewards that we are not expecting. Schultz and colleagues conducted experiments in the 1990s that showed that the release of the neurotransmitter dopamine in the reward system of the brain is critical to this experience of unexpected pleasures [98]. Dopamine is released when there is a big difference between the reward we are expecting and the reward we actually get. More dopamine is released to unexpected rewards. Imaging experiments have shown that this dopamine release is associated with more neural activity in the medial OFC and the nucleus accumbens for unexpected rewards [99].

Beyond our expectations, the context in which we eat and drink can have a profound effect on our experiences. Samuel McClure and his colleagues first reported how this effect works in the brain [100]. People claim to strongly prefer either Coke or Pepsi, despite the fact that both drinks are quite similar to each other. When people were given sips of Coke and Pepsi without knowing which they were tasting, whichever cola they preferred at the time produced more activity in the medial OFC and ventromedial frontal areas. They were then given Coke in two conditions. In one, they didn't know what they were drinking, and in the other, the cola was labeled "Coke." Once they were told the brand of the cola they were drinking, the medial OFC got activated, and this activity was accompanied by additional activity in the hippocampus, midbrain, and dorsolateral prefrontal cortex. Activity in these additional areas was presumably driven by memories of and knowledge about their past experiences with Coke, and that knowledge modified the neural response in medial OFC. The point is that the people were not fooling themselves about experiencing pleasure; knowledge actually changed their experience of pleasure.

Other experiments confirm and extend the contextual effects of pleasure reported by McClure and colleagues. Odors labeled as cheddar cheese are rated as pleasant and activate the medial OFC and ventromedial prefrontal cortex but not when the same smells are labeled as body odor [101]. Similarly, people prefer wines that they think are expensive. This preference is accompanied by more activity in medial OFC [102]. These contextual effects are not just confined to the laboratory. People at dinner will rate the same wine as better when it is in a bottle labeled from California than when it is labeled from North Dakota [103]. Interestingly, diners eat more food when they think they are drinking a finer wine, as if the pleasure from one overflows into the other.

You would think that the pleasure and pain of food and drink would be organized rigidly. We saw that smells and taste have direct access to our reward systems (see Figure 2.1). They are not filtered through the thalamus like other sensations. They arrive in parts of the brain that are not close to language areas, making them hard to describe. As an aside, this anatomical organization is probably why words used by wine experts in describing wines can be so bewildering to the average person. If smells and tastes have unfiltered access to the reward system of our brain, are difficult to corral with language, and are critical to our survival, you would expect that this system would not have much room for flexibility. Yet, as we saw, our culinary pleasures are deeply influenced by our past experiences, by what we expect, and by what we think we know. If our pleasures from smell and taste are so modifiable, imagine how flexible other pleasures might be, especially for objects far removed from basic survival, like art.

Foods can target our pleasures exquisitely. Chocolate is the quintessential example of something that for centuries has honed in on our pleasure centers. The Olmec and Mayan civilizations consumed chocolate 2,600 years ago [104]. Spanish conquistadors raved in their journals about Mayan preparations of chocolate with dried ground cacao beans, water, honey, and chili peppers. Chocolate can drive people to great lengths to get their fix. Linneaus, the great biologist, called chocolate "theobroma cacao," or food of the gods.

Chocolate is good for us. People that eat 1.4 ounces of chocolate a day for 2 weeks produce less stress hormones [105]. Chocolate contains over 350 compounds that work through three major neurotransmitter systems

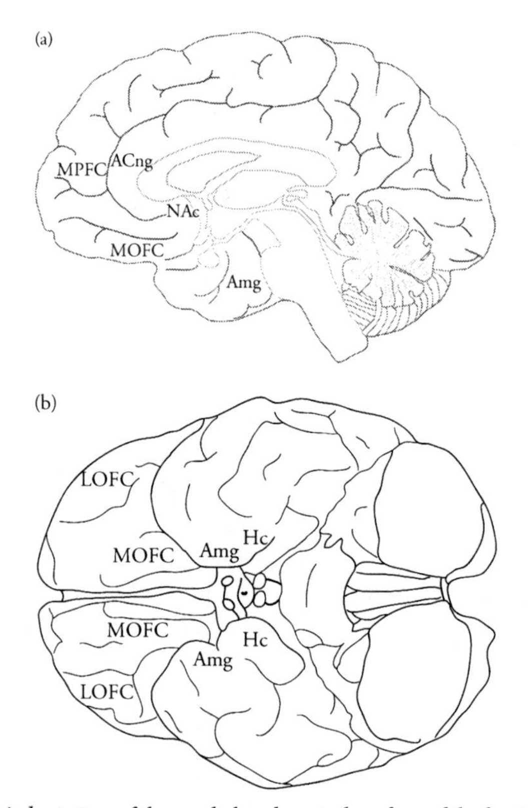

Figure 2.1. A depiction of the medial and ventral surface of the brain. The areas labeled are important for pleasure and reward systems. Abbreviations: ACng, anterior cingulate; Amg, amygdala; Hc, hippocampus; LOFC, lateral orbitofrontal cortex; MOFC, medial orbitofrontal cortex; MPFC, medial prefrontal cortex; NAc, nucleus accumbens.

(dopamine, opiate, and cannabinoid) that are the basis of our reward chemistry. First and most basic of the ingredients in chocolate is sugar. We are designed to seek sugar as a source of nourishment. Sugar calms us, an effect seen most obviously when a sweet liquid is placed on the tongue of a crying newborn [106]. Sugar activates the opioid system in the brain, which relieves stress. Chocolate contains two mild stimulants, theobromine and phenylethylamine, that affect the dopamine and norepinephrine systems and increase our arousal. Finally, chocolate also contains compounds that are similar to natural anandamides. Anandamides are brain neurotransmitters that are named after "anand," the Sanskrit and Hindi word for bliss. Anandamides work by binding to cannabinoid receptors [107]. Marijuana has its effects through these receptors. The bliss that people can feel while eating chocolate is produced through this system. Depressed people eat more chocolate than people who are not depressed. It makes perfectly good sense that people who are depressed crave something that relieves their stress, wakes them up, and makes them mildly blissful.

Foods other than chocolate target our pleasure centers and produce cravings, sometimes with unfortunate consequences. A few years ago I went to Botswana to work at a public hospital. My hospital at the University of Pennsylvania has a program in which physicians, medical students, and other health care workers spend time at the Princess Marina Hospital in Gaborone, the capital of Botswana. During one of the orientation sessions, I met a social scientist who was going there to study the epidemiology of obesity. I couldn't help but think she was joking. A quarter of the people in Botswana at the time were infected with HIV. This was a nation hobbled by "the wasting disease." Studying obesity in Botswana sounded like going to the desert to study the effects of flooding. I was wrong. Candy and cookies are swamping the traditional diet of vegetables with pap (a traditional porridge made of ground maize) and Seswaa (meat simmered slowly in salted water). Stalls, mobbed by children, dole out sweets at every schoolyard entrance. Western multinational corporations flood the streets and the children of Gaborone with junk food. The very real fear is that in a generation or two, the major medical epidemic in Sub-Saharan Africa will shift from AIDS to obesity, with its entourage of diabetes and cardiovascular and cerebrovascular disease. This is one domain in which this region of the world may soon catch up with the United States. A third of Americans are obese and another third are overweight, and the numbers are expanding.

Why do we have an epidemic of obesity in the United States? Certainly our increasingly sedentary lifestyles contribute to widening waistlines. Genetic factors also play a role. Furthermore, in stressful times, people turn to food for solace. Good-tasting foods produce beta-endorphins that counteract stress. However, another important reason for our obesity is that our foods, like our missiles, have become smart. We risk being demolished by our food.

Junk food exploits our cravings for sugar and fat. Our Pleistocene ancestors needed sugar and fat as energy sources to survive. These resources were in short supply, and those who found pleasure in high-energy foods, such as fruit and mother's milk, had an adaptive advantage. The desire for sweet as an immediate energy source seems obvious. But why fat? Dietary fats, particularly the omega family of fats, are critical to human babies' brain development. Natural stores of these necessary fats last for about 3 months after birth, so human infants need to get these fats from their diet for their brains to develop normally. Michael Crawford and colleagues of the Institute of Brain Chemistry in London speculate that our brains could only evolve to their current size after early hominids moved to shoreline environments where omega fats from seafood were plentiful [108]. Ratings of how much people like food are directly related to proportions of sucrose and fat in the food [109]. For example, the fatty acid linoleic acid increases the neural response of taste buds to sweet, salty, and sour tastes [110]. Natural preferences or aversions are enhanced by fat.

Our approach to food, which was critical to survival tens and hundreds of thousands of years ago, no longer applies to modern lives, at least in parts of the world where food is plentiful. In much of the developed world, high-energy and low-cost food is easily available. Junk food capitalizes on our evolutionary predispositions for sugar and fat. It works using the peak shift principle that we encountered with responses to beauty. That is, take a stimulus that produces a certain response and then exaggerate its critical features to exaggerate the response. Junk food is a caricature of the real nourishment we seek. It evokes the exaggerated response of overeating.

Nora Volkow, of the National Institutes of Drug Addiction, and her colleagues argue that behaviors that lead to obesity and drug addiction are similar [111]. The same opioid, cannabinoid and dopamine receptors, propel these addictive behaviors. Animal studies show that brain regions involved in food craving include the amygdala, anterior cingulate, OFC, insula, hippocampus, caudate, and dorsolateral prefrontal cortex. These are the same areas involved in drug addiction.

In summary, food serves as an excellent starting point to think about our pleasures. Food is obviously and directly tied to our survival. Our biology has been sculpted by evolution to prefer objects we need to develop normally and sustain ourselves. The experience of flavor is a result of a complex integration of different kinds of inputs. We have mechanisms that enable us to either anticipate (smell) pleasures of food or to experience (taste) them in the moment. Pleasures from food help us make new associations. Different experiences of pleasures engage similar areas in the brain. The feelings of pleasure are modified by our experiences and expectations. Finally, based on the peak shift principle and the fact that our brains evolved in environments unlike where we find ourselves today, our pleasures can hurt us. Junk food capitalizes on our adaptive desires for sugar and fat.

As we shall see next, many of the principles that underlie our pleasure in food apply to our other appetitive pleasure, sex.

Chapter 3

Sex

Before I realized what was happening, the patient reached down between my legs and grabbed my genitals. It was 1985, in the middle of the night during my medicine internship. I was working about 110 hours a week. Every third night I was on call and felt lucky if I got a couple of hours of sleep. That night, I was taking care of this patient for another intern. On my endless "to do" list was the task of placing an intravenous line. When I got to her room it was dark. I didn't know what her medical condition was. I was focused on starting her IV and then moving on to my next task. I turned on the soft light over her hospital bed and gently woke her. She seemed calm. I loosened her restrained arm to look for a good vein. That was when she grabbed me.

Even in my sleep-, food-, and sex-deprived state, I recognized that my charms were not the reason for her attention. She acted indiscriminately. She grabbed nurse's breasts and students' buttocks with the same enthusiasm. I had not yet started my neurology residency and did not know that she was suffering from a human version of Klüver-Bucy syndrome. The syndrome is named after Heinrich Klüver, a psychologist, and Paul Bucy, a neurosurgeon, who observed that rhesus monkeys changed profoundly when their anterior-medial temporal lobes were removed [112]. They became placid. They were no longer fearful of objects they would normally avoid. They became "hyper-oral," meaning they would put anything and everything in their mouth. They also became hypersexual. A similar syndrome occurs in humans [113]. The patient I encountered that night had an infection affecting parts of her brain analogous to those parts in monkeys that Paul Bucy removed. All the cultural and neural machinery that puts a check on such behavior was dissolved by her infection. She displayed sexual desire, the deep-rooted instinct that ensures the survival of our species, in its most uninhibited form.

People are preoccupied by sex. In an American national survey from the mid-1990s, over half the men and a fifth of the women reported thinking about sex at least once a day [114]. In an earlier survey from the 1970s, people were called at different times of the day and asked if they had thought of sex in the last 5 minutes. For people between 26 and 55, 26% of men and 14% of women said yes [115]. Sex sells. Pornography is one area that has not had trouble surviving commercially on the Internet. By some accounts, over $3,000 dollars are spent every second on Web-based pornography [116]. Lest you think this propensity is a peculiarly human obsession, it turns out that male rhesus monkeys also watch pornography. Researchers at Duke University found that male monkeys choose to watch pictures of aroused female monkey behinds even if it means foregoing juice rewards [117]. As an aside, they also look a lot at high-status males. Our preoccupation with sex and power is built into the hardware of our simian brains.

Despite the fact that sex takes up a huge part of our cognitive and emotional mental space, scientific research on sex has been limited. Breakthrough reports, such as those by Kinsey or Master's and Johnson, remain unusual. Perhaps overzealous notions of propriety have historically inhibited such research and investigators are easily branded as perverts. Recently, Ogi Ogas and Sai Gaddam examined sexual desires on the basis of search terms that more than 2 million people use on the Internet. In their sample of 400 million, more than a quarter of all search terms were about sex. Their fascinating book, *A Billion Wicked Thoughts*, generated controversy, ranging from being heralded as providing new and unprecedented insights into human sexuality to reifying cultural stereotypes and simplifying gender differences in sexual desire. Despite wariness in this research, knowledge about the neurobiology of sex is growing. Some themes that emerge from this research will be familiar from our rumination on food.

We can think of sex as a play with different acts. The first act is desire, the next one is sexual stimulation and pleasure, and the final act is the aftermath, the languorous glow of the sexually sated. Most of what we know about how the brain responds in these acts comes from studies of young heterosexual men. These specimens are found in abundance on college campuses and are quite willing to volunteer for sex studies.

We approach things that we desire and, as we saw before, the amygdala helps us do that. In the last chapter, we saw that the amygdala plays this role in our approach to food, and it seems to be true for sex as well. In animals, the amygdala activates their sexual response, a pattern also seen in humans. When young men look at short arousing video clips, their amygdalas are active [118]. We think such activity arouses them to move toward objects of their desire. After a successful approach, when the penis or clitorus is stimulated, the amygdala becomes less active. Thus, amygdala activation is critical in getting us to act on our desires and then settles down when we receive them.

The neurotransmitter dopamine plays an important role in our desires. The brainstem sends dopamine to many areas of our reward systems, like the ventral striatum (especially its major subcomponent, the nucleus accumbens), the amygdala, the hypothalamus, the septum, and the olfactory tubercle. As we saw earlier, these areas are involved when we desire food. They are also involved when we desire sex. The neuroscientist Itzhak Aharon and his colleagues showed that heterosexual men will exert extra effort to view pictures of attractive women, and that this effort is associated with more neural activity in the nucleus accumbens [48]. Cocaine and amphetamines amplify the effects of dopamine and enhance the desire for sex. Neural activity in the hypothalamus that increases during sexual arousal is enhanced with the drug apomorphine, which works on dopamine receptors [119]. Conversely, antipsychotic medications and some antidepressants that block dopamine receptors inhibit sexual desire.

Dopamine lets us anticipate sex but does not itself cause the intense peak of sexual pleasure. When men with erectile dysfunction are given apomorphine, they have more neural activity in their brains in response to sexually arousing images, without increasing their pleasure [120]. Neuroscientists can conduct studies of how dopamine regulates the anticipation of sex in rats with a detail not possible in humans. By inserting very small catheters, they measure the chemical environment in areas important for rewards. When a male rat is separated from a receptive female by a barrier, his nucleus accumbens is flooded by dopamine. If the male rat is then allowed to copulate with the female rat, dopamine levels plummet.

However, if the rat then sees a new female, his arousal and the dopamine levels rise again [121].

Given how engaging sexual experiences can be, it is no surprise that many parts of the brain are active when people are sexually aroused [122]. The insula, the anterior cingulate, and the hypothalamus get into the act. The insula monitors the internal state of the body and regulates our autonomic nervous system, including heart rate, blood pressure, and sweat responses. The anterior cingulate monitors for mistakes to guide future behavior. The hypothalamus regulates the secretion of hormones such as prolactin and oxytocin into our bloodstream. In addition to the usual reward systems, parts of sensory cortex also get engaged.

As you can imagine, it is hard to study what happens in the brain during orgasm. From the little information we have, the ventral striatum is active in men and in women. That activity is to be expected, since so many studies link the nucleus accumbens, a major subcomponent of the ventral striatum, to pleasure. Interestingly, activity in many parts of the brain decreases during orgasm [123]. The ventromedial prefrontal cortex, the anterior cingulate, the parahippocampal gyrus, and the poles of the temporal lobes decrease their activity. The ventromedial prefrontal cortex is engaged when we think about ourselves and about our fears. The anterior cingulate is engaged when we monitor mistakes. The ends of the temporal lobes organize our knowledge of the world, and as we saw in the discussion of landscapes, the parahippocampus represents our external environment. What could a drop in neural activity in these areas mean? Perhaps it means that the person is in a state without fear and without thought of themselves or their future plans. They are not thinking about anything in particular and are in a state in which the very boundaries that separate them from their environment have disappeared. This pattern of deactivation could be the brain state of a purely transcendent experience enveloping a core experience of pleasure.

In French literature, the release from orgasm is famously referred to as *la petite mort,* the little death. Freud thought that orgasms opened the way for Thanatos (the death instinct) after Eros had departed. These death images capture the lassitude that follows orgasm, but not the emotionally satisfied feeling. The satisfied state probably results from release of a combination of beta-endorphins, prolactin, and oxytocin. The hypothalamus regulates the production of prolactin and oxytocin. Prolactin, a hormone that helps women produce milk when breastfeeding, contributes to the sense of sexual satiety. At least in men, prolactin plays an important role in the refractory period after orgasm during which men have little further sexual desire. Given the blockbuster sales to men of drugs like Viagra, it is no surprise that prolactin-inhibiting drugs are being researched with the hope of minimizing this refractory period. Oxytocin is a hormone associated with trust and a sense of affiliation [124]. In sex, it is the "cuddling" hormone. Users of the death metaphor for the post-orgasmic state simply ignore the warm glow of endorphins and oxytocin, unless they know something about death that the rest of us do not.

When people are sexually satisfied, they have more neural activity in the lateral OFC [123]. This is the same pattern of increased neural activity seen in people who are sated with food. Neural activity in this area suppresses our reflexive tendency to act on urges. Damage to this area as well as to the anterior and medial temporal lobe can produce hypersexuality. These areas that regulate behavior, either because desires have been satisfied or because acting on desires could get us into trouble, were almost certainly damaged in the patient that made a grab for me.

Pleasures are more than simple reflexive reactions to desirable things. We saw this principle with food, and the same applies to sex. The context in which we encounter objects makes a big difference in our subjective experiences. For example, pain can topple into pleasure. Women have higher thresholds for pain when sexually aroused. These thresholds increase on average by 40% with vaginal stimulation and by 100% near and during orgasm [120]. Despite these changes in what counts as pain, the sensation itself is not dulled and is no less arousing. Rather, the same intense sensation is not experienced as pain. In the brain, the insula and anterior cingulate are active during arousal [125]. These same areas are active when people feel pain. Curiously, people's faces take on similar contortions when experiencing intense pain as when experiencing orgasms. Here the sensations producing pain are still experienced, but they are not unpleasant.

Why should brains have a mechanism to keep the arousing properties of pain and discard their unpleasant ones? The adaptive significance of this mechanism is probably to reframe the pain of childbirth. Minimizing pain during the "vaginal stimulation" of childbirth is a good thing if women are to repeat the event. This adaptive mechanism explains why otherwise painful stimulation can be pleasurable during sex. The sensations remain intense and during sexual arousal are not aversive. An adaptive mechanism that evolved for procreation got co-opted for recreation.

Pleasures help us learn. In animals, food or juices are commonly used as rewards. In the same way that food can be paired with something neutral to make Pavlov's dogs salivate to bells and whistles, sex can be associated with neutral objects. This association is one way that fetishes develop. In the 1960s, researchers exposed young men to sexually arousing images along with knee-high boots. After the exposure, these men found boots sexually arousing [126]. Linking sex to neutral things may be especially powerful during adolescence when our brains and behaviors are being molded by sex hormones. This phenomenon explains in part why fetishes can seem bizarre to people who do not share the fetish. It is the intrinsic neutrality of the fetish object that makes it seem so strange if you have not had the experience of pairing it with the pleasure of sex.

The use of sexual pleasure for learning has a dark side. The annals of medical therapy include the use of this kind of learning for deeply disturbing purposes. The episode that I am about to recount is a detour from the main points of this chapter, but I feel compelled to tell it, perhaps as a confessional in shame for my profession. *Anhedonia* is a medical word for the lack of pleasure. It is a common symptom in mental illnesses like depression and schizophrenia. In the 1950s and 60s, researchers were making great strides in mapping the neural bases of emotions. They discovered that electrically or chemically activating deep parts of the limbic system produced intense pleasure. The researchers were probably stimulating the nucleus accumbens. In people, such stimulation produced multiple orgasms. Robert Heath, a psychiatrist, worked with these stimulation techniques to alleviate anhedonia in patients. He was an early advocate of biological psychiatry, believing that most psychiatric illness had a physical basis, before this was a popular idea. He also thought that the stimulation technique could treat homosexuality.

In 1972, Heath published a study with Charles Moan [127] that used deep brain stimulation in a man referred to as B-19. This 24-year-old man had a troubled psychological and social background. His father was abusive and drank excessively. His mother was withdrawn and rigid. B-19 had no memory of ever being embraced by her. He was expelled from schools three times by the age of 11. He then dropped out of school and had a few short-lived jobs. Then he enlisted, but was discharged because of "homosexual tendencies." He was described as being hypochondriacal and paranoid. He became addicted to alcohol and drugs, but said that he did not receive pleasure from them or from sex. Heath and his team placed electrodes throughout B-19's brain, including frontal, parietal, septal, and hippocampal regions. Only electrical activity in the deep limbic regions produced pleasure. Dr. Heath saw an opportunity to "cure" homosexuality, which at the time was labeled as a disease by the American psychiatry establishment. B-19 was shown 15-minute "stag" films of a man and woman having sex while his brain was stimulated. To "test" the effectiveness of his treatment, a 21-year-old prostitute was brought to his room. B19 was able to have sex with her. After this treatment, he had a short-lived affair with a married woman. He continued to have sex with men because (according to the researchers' report) hustling was a quick way to make money. However, the doctors concluded that an important part of the study was the "effectiveness of pleasurable stimulation in the development of new and more adaptive sexual behavior." The next year, in 1973, homosexuality was removed from the list of diseases in the *Diagnostic and Statistical Manual of Mental Disorders*. To my knowledge, studies such as those done with B-19 did not continue.

What can we say about sex and pleasure? Clearly, sexual pleasure is adaptive in the most basic of ways. Enjoyment in sex guaranteed that our Pleistocene ancestors begat us. They did not have the option of making babies in a lab. This pleasure system, like that of food, has components of desire, components of actions to satisfy those desires, and components that revel in pleasure itself. There are systems that put breaks on our sexual behavior. Pleasures help us learn and develop emotional bonds to objects that are not inherently pleasurable. Finally, the pleasure of sex can change depending on its context. Painful things can become pleasurable and pleasurable things can become painful if doused with guilt and shame. Like food, the basic pleasure of sex is malleable. The fact that these experiences are so supple is critical to understand when we consider our responses to beauty and to art. Aesthetic encounters too can change radically depending on the context and the experiences we bring to the encounter.

Pleasures are promiscuous. The fetish example shows us that pleasures attach easily to other objects. These other objects include money. Some time ago, I was eating a fine dinner at an upscale Italian restaurant in West Palm Beach, Florida. West Palm Beach is one of the richest communities in the United States. I was there as part of a fund-raising effort on behalf of the University of Pennsylvania's School of Medicine. A few professors gave brief talks about science to an ultra-rich audience in the hopes that they would feel good about learning and then feel good about writing big checks to support the institution. This fancy dinner topped off the event. To my left was a dapper man in his late 70s. His date, a woman about 25 years younger, was wearing jewelry that oozed money. My dapper companion turned out to be a charming conversationalist. Our discussion took off when he found out about my interests in aesthetics. He talked about his own dabbling with painting over the years and past interactions with the painter Fernand Léger. I mentioned to him that I was planning to write a book on the science of aesthetics. As he listened to me, coddled by food and wine, he drew me in close to share his wisdom. "If you want the book to sell," he said, "make sure you include a lot of sex." This chapter and the transition to the next one about money is in honor of my dapper dinner companion. He knew that sex is tied closely to money. How closely? That is one of the topics of the next chapter.

Chapter 4

Money

"Money, money, money, money, money!" the man exclaimed excitedly. His entire vocabulary consisted of this one word. When angry, he spat "money" out as if it were a curse. When fearful, or happy, or sad, he expressed his emotions through the same word. This man was a patient on our neurology ward at the Hospital of the University of Pennsylvania in the late 1990s. He had suffered a large stroke in his left hemisphere, rendering him virtually speechless. He was reminiscent of one of the most famous patients in neurology, the case of Tan. Tan, reported by the neurologist Paul Broca in 1861 [128], was really named Leborgne. At the time he was called "Tan" because it was the only syllable he could utter. He had also suffered from a large stroke affecting the left hemisphere and was the sentinel case pointing to the lateralization of language in humans. Like our money man, Tan expressed a wide range of emotions using the one word at his disposal. Since 1861, many cases like Tan have been observed. Patients may have one or two words, or they may utter meaningless syllables. Patients like Leborgne and our patient lose access to most of their vocabulary. But some words survive, resistant to the ravages of their brain damage. I have no idea if our patient was a banker or financier, but "money" was embedded deep in his brain. While our patient was an extreme case, money is often embedded deeply in our brains.

Why should I consider money in this discussion of pleasure? Money is not directly tied to our appetites. We do not usually eat money or have sex with it. It seems like an abstract object. Historically, economists assumed that we are rational when we make monetary decisions, that we maximize our benefits and minimize our costs. They also assumed that people know what they like and that these preferences remain stable. In this view of the world, our decisions are fueled by logic and carried out with deliberation.

We identify the relevant information and analyze situations accurately. If this were an accurate description of the way we relate to money, then we might only be interested in money as a yardstick for pleasure. Money might simply be a way to quantify how much we are willing to pay for the pleasure of food or sex.

Our relationship to money, however, is far more messy and interesting, and it is relevant to our discussion of pleasure. Most of us are not rational about money. This irrationality comes from at least two sources. First, many of our decisions are more automatic than we like to think. We use quick and dirty shortcuts to make many decisions. These shortcuts probably evolved because they were useful to our ancestors. Second, our decisions are often colored by emotions, both positive and negative.

We enjoy receiving money and feel pain when we lose it. These experiences are why money gives us a context in which to think about receiving pleasure from something removed from our immediate appetites. Do we find pleasure in money itself? Are the neural underpinnings of pleasure from money similar to the pleasure we get from food and sex? Money also provides us with a way to explore in greater detail how and why we distance ourselves from pleasure. As we saw with food and sex, there are situations in which "approach behaviors" need to be tempered. Money offers many situations in which we choose between immediate and later rewards.

The field of neuroeconomics is now in high fashion. Scientists are optimistic that knowing how the brain works will tell us important facts about how we make decisions about money and hopefully guide us to make better ones. The first question relevant to our discussion is, does money itself give us pleasure? Why should it? Money is simply paper or metal or numbers on a bank record. If I were hanging a picture on a wall and I picked up a hammer, it would not be obvious that the hammer gives me pleasure. The hammer is useful and I take advantage of its utility. If my brain were scanned when grabbing the hammer, the neurons in my reward centers would probably show that they don't care. The situation differs with money. Money activates our reward system.

Money, like sex, engages brain systems that anticipate the pleasure of receiving money and systems that give us the experience of pleasure when we actually receive money. In fact, money is such a powerful reward that it activates part of our reward system even when we are subliminally exposed to pictures of money [129]. When people receive real money in laboratory economic games, parts of the ventral striatum and medial orbitofrontal cortex (OFC) get active. Even the anticipation of money activates the ventral striatum. By contrast, the medial prefrontal cortex activity seems to keep track of money after it has been received [130].

Our pleasures are also tied to the absence of pain and loss. In the brain, pain and loss are not just less activity in pleasure areas. Rather, as we saw with food and sex, other brain structures actively code pain and aversion. These structures fire when we lose money [131]. Parts of the brain that code aversion, such as the anterior insula, the lateral OFC, and parts of the amygdala, are active when we experience risk and uncertainty [132]. As we saw earlier, the lateral OFC is more active when we are sated by food and drink. In that situation, the same taste becomes less pleasurable, sometimes even aversive. The anterior insula is tied to our autonomic nervous system and is active when we experience disgust. This activation is the brain correlate of the "ick response," our visceral disgust for things like rancid food. The same structure seems to be active when we experience disgust in financial transactions.

The neural circuitry that anticipates pleasure is distinct from the circuitry that codes the experience of pleasure, which is also distinct from the circuitry that makes us choose to act on our desires. Whether the anticipation, enjoyment, and choice of actions engage the same respective neural structures when faced with different kinds of rewards, like food, sex, or money, is not clear. Most studies show activation in the medial OFC when people experience pleasure from different sources. In a study conducted in Lyon, France, neuroscientists found that the ventral striatum, anterior insula, anterior cingulate cortex, and midbrain encode the subjective reward regardless of the type (in this case, money and sexually provocative images), which suggests that the pleasure of money and sex are coded similarly in the brain. But in a different part of the OFC, they found that monetary gains produced activity in an evolutionarily newer part of the brain and erotic images produced activity in an evolutionarily older part [133]. Perhaps part of the pleasure we derive from money is a later development coded in brain regions present in most primates but not in earlier mammals.

Why would money activate areas of brain similar to those that are active when we receive the most fundamental of pleasures from food and sex? When I was a student at Haverford College in the late 1970s, my friends and I played pinball at night. We often played after we had spent most of the evening studying in the library. There was typically a scramble for the quarters needed to play. At some point, we started calling quarters "pleasure disks." They were little pieces of pleasure. Most of us experience money paired with pleasures so often and so early in our lives that it takes on the mantle of pleasure itself. This linking of money to pleasure is similar to tones paired with food in Pavlov's dogs or the knee-high boots that get fetishized when paired with sexually provocative images. The Merriam-Webster's dictionary defines a fetish as something believed to have magical powers, or an object of irrational reverence or obsessive devotion. It strikes me that as a culture we (in the United States) fetishize money.

Daniel Kahneman and Amos Tversky, both psychologists, profoundly influenced how we think about economics [134]. Kahneman won the Nobel Prize in economics, a prize Tversky missed sharing because of his premature death. They heralded the field of behavioral economics, which has uncovered many situations in which we act in less than logical ways. Humans have all sorts of biases that color decisions, including financial ones. To illustrate the dynamics of financial decisions, I will look at institutions that are master money manipulators. I do not mean the too-big-to-fail financial firms on Wall Street. I mean the impossible-to-fail casinos in Las Vegas and Atlantic City.

How we frame a situation has a profound effect on our emotional coloring of that experience [134]. Casinos and most advertising agencies appreciate the power of framing our choices, even when the actual information provided is exactly the same. For example, people might be more likely to buy a raffle ticket that has a 10% chance of winning than one that has a 90% chance of losing. Casino ads emphasize the chances of winning without ever mentioning the greater chances of losing. Casino packages emphasize rewards for staying in their resorts and the various discounts we can receive, as if they were giving us something, when in fact we are paying for their hospitality.

Developers of casinos understand that we are social creatures. Our satisfaction is largely determined by how we stack up relative to others. People make seemingly irrational choices when given the following two scenarios. Imagine being in line, with a chance of receiving cash when you reach a counter. In one scenario, you get $100. In another, you get $150, but the person ahead of you gets $1000. Most people prefer to be in the first scenario even though they end up with less money than in the second one. Once our basic needs are met, we care more about our relative position in a group than in some absolute measure of our reward. Recognizing this fact, casinos are designed to segregate players based on the kinds of bets they make. For the casino, it makes no sense to have someone who wins $150 be unhappy because the person at the next table just won $1000.

Casinos also minimize the effects of what social psychologists call "endowment." Endowment means that we endow objects we have with more value than if we didn't possess the same object. Casinos want us to devalue our own money so we are not so bothered by losing it. To understand the endowment effect, consider the following situation. Some people are divided into two groups. One group gets a free mug and the other a free pen. Both items are worth the same amount of money. The two groups are then given the option of exchanging their gifts. Since some people might prefer to have a mug and others a pen, one would expect that a lot of people would trade. However, very few do [135]. Just having something in our possession enhances its value. In some studies, people value what they own as up to twice what buyers are willing to pay. Money plays on this effect. If it is in my pocket, it is mine. My guess is that the endowment effect is even greater for distinctive objects like the currency design for a given country. Our paper bills are designed with great care and are wonderful aesthetic objects. Many people find the currency of their own country more beautiful than those of other countries. Americans often regard the different colors of the bills from other countries as garish or play money, whereas people from elsewhere regard U.S. dollar bills as boring, with each denomination being the same size and color. We endow cash, as an aesthetic object in our possession rather than as a symbolic means of exchange, with a value inflated beyond the goods and services it buys. We are reluctant to give it away and feel pain when we do.

We hate losing. My partner, Lisa Santer, does not like playing competitive games. She is very smart and competitive and actually wins most games that she plays. But, more than liking to win, she hates losing. So she avoids playing if there is more than a small chance that she will lose. We all have varying degrees of this aversion to loss built into our brains, whether it is when competing or exchanging goods. On average, people dislike losing, twice as much as they like winning. This bias makes most people inherently conservative. The more we value something, the more we hate losing it.

Casinos, and companies in general, go to great lengths to disguise the pain we feel with losing money. When we pay with cash, we concretely give away something that we possessed. When we pay with a credit card, the card comes back to us. We are willing to pay more for something when paying by credit card than with cash [136]. Although there are financial reasons to prefer paying by credit card, the financial advantage of using the card over cash does not explain how much more people are willing to spend with a credit card.

There are many examples of people paying more for services when the costs are hidden. People oversubscribe to flat-rate payment plans for utilities and telephone service and fitness clubs. A flat-rate plan allows people to enjoy the service without thinking about additional costs. People also like package deals in which components are presented as "free," even though the claim is meaningless. Casinos and resorts capitalize on all-inclusive packages, creating the illusion that they are giving something away, when consumers are the ones relinquishing their money. In this context, ad-hoc abstract currencies such as frequent-flyer miles or play money like beads used at vacation resorts hide the pain of payment. Casino chips serve the same purpose. Making chips relatively uninteresting and indistinguishable minimizes their endowment effect. We don't feel like they belong to us in the same way as our paper currency. With this lowered sense of value, we feel less pain when we end the night with fewer chips than we started with.

Companies have become so good at hiding costs that financial institutions even manage to hide the costs from themselves. The big financial meltdown of 2008 was driven by the fact that complicated financial derivatives hid the risk of losses, making it easier for Wall Street to indulge in increasingly risky behavior. Like addicts, Wall Street craved rewards without regard to risk. There was a free lunch, until there wasn't.

Casino environments are designed to encourage us to make less than rational decisions. Many neuroscientists think that most of our decisions boil down to three types. These decisions are Pavlovian, habitual, and goal-directed. Importantly, these decisions can work together to reinforce our behavior or work at odds to place us in quandaries. Pavlovian decisions are reflexive and automatic. Habitual decisions are complex and learned, and over time they also become automatic. Goal-directed decisions are also complex, but are made with thought and deliberation.

The details of these systems and how they interact are debated among scientists [137]. However, the fact that we respond automatically or deliberately and are driven by emotions or logic is clear. Casinos want us to make automatic and emotional decisions rather than make deliberate and logical ones. They want us to keep playing and lose money, while keeping hope alive that we will win even when we know better.

"All your desires, all in one place" proclaims Harrah's casino Web site. Casinos provide plenty of basic rewards. They plow us with cheap and plentiful food. Gambling areas are dotted with scantily clad attendants ready to lubricate us with alcohol. Sex workers are no strangers in these places. Casinos foster Pavlovian reflexive and automatic responses to objects in the environment that give us pleasure. Pavlovian choices are expressed in the brain through the amygdala [138] and its connections to the nucleus accumbens, and the hypothalamus. As we have seen before, these structures reside deep in the brain, and we are usually not aware of their influence on our behavior. As a result, we are often not even aware of our decisions being driven by Pavlovian responses.

Casinos encourage habitual behaviors by linking them to Pavlovian ones. Habitual behaviors, as the name implies, come from repeated behaviors that we learn over time. For example, when we drive to work on a route that we take every day, we make all kinds of decisions, such as whether to accelerate or brake, or whether to turn or go straight. These decisions have an automatic quality and usually hover at the edge of our consciousness. The actions take time to learn, but once learned, they persist.

Pavlovian and habitual behaviors interact, sometimes in harmful and sometimes in helpful ways. People with obsessive-compulsive disorders feel compelled to repeat rituals. Here the repeated behaviors decrease their anxiety and provide mild pleasure. In other situations, environments are designed to strengthen links between Pavlovian and habitual behavior. Anybody who has watched people at slot machines knows exactly what I mean. Watching someone pull the lever of several machines repeatedly has all the characteristics of a ritualized habit. This lever-pulling habit is paired with rewards in the form of unpredictable and noisy cash deliveries. Slot machines pair habitual and Pavlovian behaviors, so that people keep playing (and paying) over and over and over. It is no wonder that we call the behavior of addicts "a habit."

We see the same pairing of Pavlovian and habitual behavior in other situations. When religious people repetitively move up and down as they daven, or whirl around a single axis with one arm raised in the air, habitual behavior is presumably being linked to the pleasure of religious ecstatic states. More mundanely, athletes combine these systems all the time. Most professional basketball players at the free throw line move in idiosyncratic but habitual ways before taking their shot. The intermittent reward of making the shot keeps the ritualized behavior going.

Even though casinos are most clearly capitalizing on the first two types of decision-making, they understand and manipulate goal-directed behaviors as well. Besides food, sex, and opportunities for obsessive-compulsive behaviors, casinos also offer the chance to win money with deliberation. Playing blackjack involves making strategic decisions. Each decision is based on information such as what cards are evident, how they compare to the player's own cards, and so on. These games engage goal-directed behaviors. Goal-directed behaviors are more flexible than the Pavlovian and habitual systems. We use goal-directed decisions to evaluate situations and change what we do based on changing circumstances in our environment. This system is easily accessible to consciousness. We can usually describe the reasons for our choices. This system engages frontoparietal circuits and activates areas such as the dorsal striatum, the insula, the anterior cingulate, and the OFC. Since some of these brain areas also encode pain and aversion, the system seems designed to take negative emotional factors into consideration when making "rational" cost–benefit decisions.

Goal-directed deliberative behavior is the closest behavior to making rational choices. However, the blackjack example also shows how rationality can fall short of guiding behavior. It is no secret that the odds in casino games are set to favor the house. Players know that if they play long enough and don't count cards (for which they are thrown out of the casino), they will lose money. Yet they play for hours on end. Why? The answer is that even in this scenario, casinos capitalize on irrational biases built into our behavior.

Casinos maximize our joy of winning even when it does not happen very often. In fact, they exploit the very fact that winning does not happen very often. B.F. Skinner studied how rewards drive behavior. He was a major proponent of behaviorism, a field of study that dominated psychology during the first half of the twentieth century. His basic idea is that we become conditioned to repeat behaviors that reward us [139]. This fact is obvious enough. However, the interesting thing about this kind of "reinforcement conditioning" is that we repeat behaviors more often if the reward is intermittent and unpredictable than if we get rewarded every time for the behavior. That is, if I receive a reward for what I do (like apply for federal research grants) intermittently and unpredictably, I'm more likely to persist in the behavior than if I am rewarded every time I do it. This "intermittent reinforcement schedule" is most effective in making rats in labs press a lever over and over and people in casinos pull a lever over and over.

Casinos want us to make short-sighted decisions. That is, they want us to behave in Pavlovian or habitual rather than goal-directed ways. Deciding between short- and long-term benefits makes up what neuroeconomists call a "discount function." This discount function describes the relative value of a reward received soon versus one received later. Would you take $10 today instead of $20 in a month? Different people balance immediate and delayed rewards differently. One of my colleagues, Joe Kable, along with New York University neuroscientist Paul Glimcher, found the neural activity that correlates with these functions in the ventral striatum and medial prefrontal cortex [140]. Many people, like medical students, can delay their rewards. But people who are impulsive tend to be drawn to immediate rewards. Casinos are designed to get us to be impulsive.

One way to make people short-sighted is to exhaust their deliberative goal-directed systems. The thoughtfulness and flexibility of our deliberation come at a cost, which is that we fatigue. The effects of this fatigue were seen in an experiment where people were asked to memorize simple two-digit sequences or more difficult seven-digit sequences. They then walked to another room for the next part of the experiment. On the way, they passed by a table with fruit salad or pieces of cake. The group that had put in the effort to memorize seven-digit sequences was more likely to grab a piece of cake than the group that did the easy task [141]. The deliberative system took a rest. Rather than consider the worth of ingesting those calories when not hungry, the Pavlovian appetitive desire for sugar and fat took over.

We become short-sighted in many situations, like when we are tired, drunk, and emotionally aroused. Craving for one object overflows into craving for another. Addicts become more impulsive when they crave drugs and money. Sexually aroused people become short-sighted about spending money. Finally, the deliberative system is designed to think about rewards in the future. Some theorists think that we imagine our emotional state in the future to guide our decisions today [142]. The reasoning goes like this: I will go to medical school and work really long hours for years so that later I will be pleased with a socially meaningful and materially comfortable life. Of course, leveraging future well-being only works if there is a future to be leveraged. Casinos want people to keep playing without regard to the future. The longer we play, the more we lose, since the odds favor the house. The longer we play strategic games, the more we fatigue. Casinos use bright lights and sounds to keep people aroused and going non-stop. Casino floors never have clocks or a clear view of the outside environment. As a result, people have little sense of the passage of time. They lose their own natural rhythms and continue to play without taking breaks. By keeping people awake and playing longer, lubricated with alcohol provided by attractive and friendly attendants, casinos encourage people to act impulsively. The brilliant slogan "What happens in Vegas, stays in Vegas" exploits this idea of no future. A visit to Las Vegas is a wonderful bubble, removed from the stresses of the rest of our life. There are no long-term consequences, because there is no long term. We give our deliberative system a rest and let our impulses go wild.

Why do we behave in ways that let casinos take our money? How could such short-sighted behavior possibly be adaptive? The answer to this question is that our environment has outgrown our paltry Pleistocene brains. Our ancestors evolved in the Pleistocene era to survive in what the evolutionary biologist Richard Dawkins called the "Middle World" [143]—that is, the physical environment between the very small and very large, between the world of cells and the universe of galaxies. Perceptually, we are tuned to objects between these extremes. We need microscopes to see tiny objects in close and telescopes to see huge objects far away.

A similar phenomenon happens with time and social complexity. We are not built to understand effects over very short or very long periods of time. This intrinsic insensitivity to long durations probably contributes to the controversies over climate change and even evolution itself. Our brains also evolved in social environments that were less complex than they are now. In the Pleistocene era, there was relatively little division of labor. There were no capital markets or accumulations of extraordinary wealth. Exchange was about food, clothing, and social favors and not about insurance, equities, and synthetic collateral-debt obligations. Most early humans lived in groups that numbered in the 10s to 100s. These groups grew into the 1,000s only in the last 10,000 years. The idea that financial decisions and consequences now involve billions of people is beyond our natural comprehension. Our Pleistocene heritage created quick-and-dirty shortcuts, making use of Pavlovian and habitual decisions, that can now seem out of place and irrational. As societies have become larger, more complex, and more anonymous, we are still accommodating biases that worked well in an earlier time, but now appear irrational.

In summary, money offers us a way to consider pleasure that is abstract, compared to that of food and sex. Money shows us how things not directly tied to our appetites can become sources of pleasure themselves. When this happens, the reward systems in our brain react to money in much the same way as they do to food and sex. As we saw with food and sex, where context can affect these primary rewards, context affects secondary rewards like money even more powerfully. Money shows us how short-term pleasures compete with long-term pleasures. It shows us how we often behave in automatic and emotional ways rather than in strictly rational and deliberate ways. Some of our behavior with money doesn't make a lot of sense until we realize that our Pleistocene puttering brain does not always keep pace with the modern world.

Chapter 5

Liking, Wanting, Learning

In 2007, Pixar studios released the movie *Ratatouille*. The movie featured Remy, a rat protagonist who spends most of his time in a kitchen. The *Wall Street Journal* wondered whether the movie could possibly succeed. We are used to cute mice, given the superstar status of Mickey Mouse and his legion of club fans. But a rat? Who wants to see a rat in a kitchen? Yet, this movie about a rat who loves good food and fights against prejudice won audiences over. The movie even garnered an Oscar for best animated feature film of the year. At one point in the movie, Remy, who desperately wants to be a chef in Paris, is cooking mushrooms over a chimney that gets hit by lightening. He tastes the charred mushrooms and exclaims, "You gotta taste this! This is...oh, it's got this kind of...mmm, it's burny, it's melty...it's not really a smoky taste. It's kind of like a certain...Pshah! It's got like this Ba-boom! Zap! kind of taste. Don't you think?" Remy is a rat that knows what he likes and knows what he wants.

Rats, it turns out, know what they like and know what they want. The neuroscientist Kent Berridge highlighted this important distinction in our reward system: the distinction between liking and wanting. *Liking* is the pleasure we get from some objects, and *wanting* is the desire we have for those objects, even when we might not be consciously aware of these tendencies. Earlier, we saw that we have brain structures that that come online while we enjoy food, sex, or money and other structures that come online when we want to get to those pleasures. Berridge and his colleagues examined these two parts of the rat's reward system in detail [144].

We know what a rat wants because the rat will press levers and suck nozzles and run mazes for objects it desires. But how could we possibly know what a rat likes as distinguished from what it wants? Surprisingly, rats, along with orangutans, chimpanzees, monkeys, and human babies, make characteristic facial expressions when they like something and when they don't. They rhythmically protrude their tongues and lick their lips when a sweet taste is placed in their mouth. They gape and shake their head side to side when given something bitter. These facial expressions are key to observing their pleasure.

Berridge and his colleagues showed that liking and wanting are not the same thing in the rat brain. Liking, the core experience of pleasure, engages the nucleus accumbens and its connections within the rest of the ventral striatum. Neural activity in these structures is driven by mu-opioid and cannabinoid receptors. Opioid receptors are involved in the pleasure that people experience when taking opiates like heroin and morphine. As I mentioned earlier in the food chapter, cannabinoids are natural brain chemicals similar to plant cannabinoids found in marijuana. Cannabinoid and opiate receptors work together to produce our pleasures. Our wants, the desire to get to objects we like, is driven by dopamine rather than opioids or cannabinoids. The ventral striatum also contains neurons that respond to dopamine and promote wanting. These neurons are interspersed with liking opioid and cannabinoid neurons.

It makes sense that liking and wanting in the brain work together. After all, we want objects that we like and we like objects that we want. However, liking and wanting can get uncoupled [145]. For example, the drug naloxone blocks the effects of opiates and diminishes the pleasure of eating without decreasing the desire to eat when hungry [146]. This effect means that liking can be turned down without changing wanting. Wanting can also be turned down while preserving liking. When dopamine cells in rats are experimentally destroyed by a toxin, the rats simply stop eating. Scientists first thought they stopped eating because they no longer got pleasure from food. However, when sweet or bitter liquids were placed in their mouths, their faces still expressed likes and dislikes [147].

In humans, as in rats, liking and wanting can be uncoupled. When humans take drugs that block dopamine they do not change how much they like objects. However, they are less inclined to act to get what they want, a state called "psychic indifference" in the clinical literature. The opposite pattern of behavior is sometimes seen in addicts. As addiction progresses, addicts are driven by greater and greater desires for what they want. They go to great lengths to get their "fix." But addicts do not necessarily experience more liking of the drug as they spiral down into their cravings.

In this discussion of liking and wanting, we have focused on parts of the reward systems that reside deep in the brain. Cortical systems interact with these deep systems. These interactions are of great importance in humans because they mean that we are not simply slaves to our desires and pleasures. Cortical systems can represent the context in which we approach our wants and enjoy our likes. They give us control over how we might act in response to objects of desire. The orbitofrontal cortex, the ventromedial prefrontal cortex, the amygdala, the anterior insula, and the anterior cingulate take stock of our pleasures and work together with dorsolateral prefrontal cortices to help us coordinate and plan strategic behaviors ultimately designed to get pleasure or even to keep them at bay when we choose to do so.

Rewards help us learn and change our behavior. We have opportunities to learn when our experiences are not quite what we expected. For example, I am really hankering for ice cream. I want to taste that creamy rich pistachio flavor melting in my mouth and flooding my brain with endogenous opioids and cannabinoids. There is a particular shop, Capogiro's, some distance away that sells amazing gelato that I know would be perfect. But a new place just opened down the block. So I go to the closer place instead. The ice cream is fine, but not as good as Capogiro's. I get the reward, but it is less than what I was expecting. My expected and my experienced pleasure don't line up. This is a situation in which I can learn to change my behavior. Next time I might not go to the new shop, or I might decide to give it another try. After all, maybe the owners of the shop are still figuring out their business and will improve their stock. Alternatively, I might just lower my expectations. So, on a day that I am tired, I decide to go back to the nearby shop. I avoid the pain of the longer walk and at the same time I adjust my expectations. If the ice cream is better than I expected, I recalibrate my expectations and once again learn from this experience. With this new knowledge, I might be more likely to return to the new shop.

How does the cycle of making predictions and learning from the experience of mismatched expectations and rewards play out in the brain? Here again, dopamine is involved when we predict future rewards and respond to errors in our expectations. Dopamine neurons increase firing over their baseline when the reward is better than expected and decrease firing when the reward is less than expected [148]. No change in the baseline firing rate suggests a close match between expected and received rewards, a situation in which there is no reason to change whatever we are doing. These are the chemical counterparts to delight when we experience unexpected pleasures, disappointment when we don't receive the pleasure we expect, and business as usual.

In humans, the ventral striatum records the mismatch between rewards and expectations. In one fMRI experiment, conducted by O'Doherty and colleagues, people were shown a blue arrowhead and then given sweet juice. After they learned this association between arrow and juice, the timing of juice delivery was varied so that sometimes the juice arrived early after the arrowhead and sometimes it didn't arrive at all. When the juice arrived unexpectedly early, the signal in the ventral striatum increased. When it did not arrive at the expected time, the signal decreased [99]. This recalibration happens fast, within 50–250 milliseconds. One implication of these findings is that we constantly adjust our expectations as we confront reality, often without even being aware that we are learning.

Our reward systems help us learn, and at the same time we feel pleasure in what we have learned. You may recall that in the chapter on math, we saw that pleasure in figuring out regularities of the natural world is part of what makes certain number configurations beautiful. We experience pleasure when we figure things out, an effect that the developmental psychologist Alison Gopnik fancifully called "explanation as orgasm" [149]. Babies purse their lips and wrinkle their brows when presented with problems that are confusing. When they figure out the answer, they smile and look radiant. Like Berridge's rats, they express pleasure in their faces, but in this case to a solution to a problem rather than to a solution with sugar. So we have this reverberating cycle of pleasure helping us learn and what we have learned giving us pleasure. These cognitive pleasures may be the reason we experience pleasure with some conceptual art. Figuring out what they mean tickles our reward systems.

One important property of our reward systems is that while we use them to drive basic appetites, like food and sex, we also use them to develop abstract notions, like fairness. We see how the brain codes fairness in what economists call ultimatum games. The game is set up with two people. One person, A, is given a certain amount of money, let's say $100. The other person, B, is given none. The game requires A to propose a split of the $100 to B. If B accepts the offer, they both keep the money. If B rejects the offer, neither gets to keep anything. Assuming that B wants to make money, B should accept anything that A offers, since it is the difference between receiving some money and receiving nothing. Most people do not act rationally in the role of B. The tipping point for accepting an offer seems to be around a 70/30 or an 80/20 split. If offered less than $30 or $20 most people reject the offer and are willing to receive nothing because they think it is unfair for A to get so much more than they do. In fMRI studies, when people are given choices they experience as unfair, their insula is active [150]. The insula is associated with feelings of disgust, especially when we smell or taste rancid food. The insula activation in ultimatum game experiments suggests that we experience disgust at unfair practices. An abstract evaluation of another person is making use of the same neural system that is engaged when we taste sour milk.

The way we establish personal reputations also makes use of basic reward systems. In one clever study, Read Montague and his colleagues scanned pairs of people simultaneously when they were playing a game that involved money exchanges [151]. One person, the investor, has $20. This person invests some portion of the $20 in a second person, the trustee. The trustee triples the money, a fact known to both players. Then the trustee decides how much to give back to the investor, who then turns around and decides how much to reinvest with the trustee. Thus, the investor can reinvest a lot of money or only a little bit. In the process of these iterations, players develop "reputations" as being generous or stingy. Changes in neural activations are seen in parts of the striatum that use dopamine as its chemical currency. The activity goes up in the trustee's brain when the investor makes a generous reciprocal offer and goes down when the investor makes a stingy offer. In this setting, the level of trust is established based on the accuracy of making a prediction about another person's rewarding behavior. We treat these interpersonal rewards in a manner similar to how we learn from basic appetites. Establishing whom we can trust uses the same brain systems that guide us in what we might eat or with whom we might sleep.

Reward systems integrate liking, wanting, and learning. Liking and wanting normally operate together. However, as stated earlier, they can uncouple. The possibility of this uncoupling is important, especially when we consider encounters with art. Our pleasures help us learn and change our behavior, and what we learn alters the pleasure we experience. Our reward system has a built-in flexibility in which cognitive and pleasure systems interact and modulate each other. Such flexibility is also important to how aesthetic encounters guide our thinking and thinking informs our aesthetic experiences. Anything can be a source of pleasure as long as it taps into reward systems embedded in our brains. Finally, even though our core reward systems evolved to serve very basic appetites, such as the desire for and pleasure in food and sex, our evaluations of abstract notions, such as fairness or the reputation of others, make use of these same systems.

Chapter 6

The Logic of Pleasure

All creatures that move, approach the things they need and avoid the things they think will cause them harm. This approach-and-avoidance behavior is the fundamental axis around which more complicated actions get organized. For humans, pleasure fuels much of our approach behavior. Pleasure profoundly affects how we live as individuals and how we evolved as a species. It is the central ingredient of a complex reward system that lets us savor food and sex, find beauty in people and places, and delight in paintings and plays. In this final chapter of the section on pleasure, let's review some of the highlights of these complex reward systems.

Our reward systems are made up of different components that are necessary for us to experience pleasure and desire. As we have seen, pleasure and desire (or liking and wanting) are not the same thing. Beyond feeling pleasure and desire, some components within our rewards system allow us to anticipate pleasure, to evaluate the pleasures we experience, and to plan actions that get us to the objects of our desires. Still other components of our reward systems modify our pleasures, restrain our approaches, and help us learn.

Our pleasure, at its most basic, is rooted in our appetites for food and sex. Our ancestors who took pleasure in nourishing food and in having sex with healthy partners were the ones who survived and propagated. We inherited their pleasures. Pleasure also helps keep us alive by serving a homeostatic function. Homeostatic functions are processes that regulate our body to stay within a narrow physiological window in which we operate at our best. If we deviate from that window, we need to be brought back in line, and pleasure encourages us to do so. When we are salt deprived, we like super-salty tastes, and when sated, we do not like sweet syrups.

The core experience of pleasure works deep in the brain through the nucleus accumbens and the rest of the ventral striatum. These structures are active regardless of whether pleasure arises from food, sex, or money. The chemical currencies in these neural structures are opioids and cannabinoids working together. The bliss of someone in the midst of an opium or marijuana high is the result of the flooding of their receptors that are bathed more gently when we experience everyday pleasures.

Desires drive us to act. The chemical currency for desire is dopamine. The brainstem sends dopamine to parts of the striatum to motivate us to act on our desires. Other parts of the brain evaluate our pleasures. The medial orbitofrontal cortex and the ventromedial frontal lobes seem to code our pleasures and are probably the brain structures that contribute to our being aware that we are experiencing pleasure. Normally, pleasure and desire work together. We want things that we like and we like things that we want. However, these components can part ways. For addicts, sometimes wanting overwhelms liking as it gets out of control. A question to which we shall return when talking about art is what would it mean to have liking without wanting?

Other neural structures guide our actions in response to our desires. The amygdala and insula play a double role. They are active when we face pleasant and unpleasant objects. These structures get us moving to approach pleasures or to avoid pain. While pleasures certainly drive much of our behavior, we are not chained to them. For example, when we need to be wary of approaching pleasures, the lateral orbitofrontal cortex, the amygdala, the insula, and the anterior cingulate cortex start firing. These structures encode satiety, anxiety, pain, and feelings of disgust. When these structures are active, we have the gut feeling that we ought to avoid the object, even if it gives us pleasure in other settings. Cortical structures such as prefrontal and parietal areas are involved in our conscious planning and deliberations. These neural structures help modify our behavior so that we can be strategic about getting pleasures, or keeping them at bay when needed.

Pleasures are malleable. In some situations, such as during sex, pain can become pleasurable. Other objects that do not typically give us pleasure can be fetishized into becoming sexually arousing. Pleasure, it would seem, can attach itself to almost any object or situation. This promiscuity of pleasure is the reason we can enjoy so many different things and the reason people vary so much in what they like. Some enjoy the burning sensation of hot peppers in their mouth; others shrink away. Some love the feel of money; others regard it as filthy lucre. Our individual histories shape our experience of pleasures. What we know, or even think we know, profoundly influences our likes and dislikes.

Dopamine, the neurotransmitter that drives our wants, also helps us learn. Every time we are delighted by unexpected pleasures or disappointed by pleasures not received, we can learn. The firing rates of dopamine neurons calibrate our expectations of future rewards. This calibration applies to such basic situations as seeking sweet juice all the way to complex situations like learning to trust other people.

Pleasure functions homeostatically at different scales. We saw that pleasure makes us act to keep our physiology functioning well. Pleasure as a way to maintain homeostasis extends from one person's desire to eat something sweet or salty all the way to our collective enjoyment of beautiful people, pretty places, and elegant proofs. Through evolution, the species brings itself in line with the environment in order to function well. When the environment changes, the species needs to change into a new state.

Our brain that evolved to face the challenges of our past doesn't always keep pace with our jet-propelled present. Our environment is changing rapidly. When we act in seemingly illogical ways, we are often observing the behavior of a brain behind its time. We intuitively understand barter exchanges, but don't understand credit default swaps. As changes in our cultural environment accelerate, the link between our adapted predispositions and our present-day actions becomes increasingly attenuated. This attenuation between our instincts and behavior is an important point to keep in mind, especially as we think about art.

Is there such a thing as aesthetic pleasure? The pleasure we derive from aesthetic experiences is rooted in but not restricted to basic appetites. The pleasure of gazing at a beautiful person or an enthralling painting is not the same as the pleasure of sugar on our tongues. Aesthetic pleasures stretch beyond basic appetitive pleasures in at least three ways. First, they extend past desires by tapping into neural systems that are biased toward liking without necessarily wanting. Second, aesthetic pleasures are nuanced and encompass admixtures of emotions more complex than simple liking. Third, aesthetic pleasures are influenced profoundly by our cognitive systems. They are colored by the experiences and knowledge we bring to aesthetic encounters. We shall return to these stretching themes when we consider art in the next section.

Looking forward, what does art have to do with beauty and pleasure? To suggest that beauty and pleasure are adaptive is analogous to saying that we have instincts for beauty and pleasure. Do we have an instinct for art? I do not think so, at least not in a simple-minded way. I suspect that art, as we experience it, has outpaced our adapted brains. The idea that art might not be an instinct does not mean that art is not integral to our lives, or that art is not profound, or that art is not a source of great joy or an expression of great sorrow. Art can be all of these. But being all these things does not make art an instinct. Art, as we encounter it today, is largely an accident. A fantastic accident, to be sure! But, like our accelerating cultural environment, I am getting ahead of myself.

PART III

ART

Chapter 1

What Is This Thing Called Art?

On May 23, 2007, the art auction house Sotheby's sold a tin can for 124,000 Euros. An artist named Piero Manzoni produced the can. This can, one of 90, was labeled in Italian, French, German, and in English "Artist's Shit Contents, 30 gr freshly preserved, produced and tinned in May 1961." Chris Ofili, a winner of the prestigious Turner Prize, made a painting of the Holy Virgin Mary using female genitals cut from pornographic magazines. Some parts of the Virgin Mary were made of elephant dung. In 1999, its exhibition at the Brooklyn Museum of Art provoked then mayor Rudolph Giuliani to exclaim, "There's nothing in the First Amendment that supports horrible and disgusting projects!" In 1987, a cibachrome photograph by Andres Serrano received an award from the Southeastern Center for Contemporary Art. The piece, called *Piss Christ,* shows a crucifix floating in an ethereal golden fluid. The fluid turns out to be the artist's urine. The late senator Jesse Helms announced, "Serrano is not an artist. He is a jerk." Gone are the days when artists ground semiprecious stones like lapis lazuli to create deep blues that represent the noble nature of Christ or when they used real gold to show divine light. Shit and piss are now materials of fine art. Conservative politicians give voice to many people's reaction of bewilderment and disgust when confronted with these pieces of art.

What is art? Unfortunately, it is not so easy to come up with a satisfying definition. In what follows, we will look at different ways of thinking about art. First, we will explore traditional views of aesthetics and art. Then, we will consider how recent philosophers have wrestled with definitions of art. It is worth underscoring the fact that aesthetics and art are not the same. They are overlapping but different ideas. Aesthetics, as generally understood, focuses on properties of objects and our emotional responses to those properties. The object need not be art per se. It could be a field of flowers just as easily as it could be Van Gogh's painting of irises. Aesthetics typically relates to the continuum of beauty to ugly. The neuroscientist Thomas Jacobsen and his colleagues had people pick words associated with aesthetics [152]. The word most commonly picked was *beauty*, over 91% of the time. The next most commonly picked word was *ugly*, at 42%. Most people think that the continuum between beauty and ugly is what aesthetics is all about. This intuition about aesthetics also pervades analytical philosophical traditions and has historically influenced theorizing about art [153]. However, aesthetic encounters need not be confined to beauty. The philosopher Frank Sibley listed examples of other aesthetic properties that include objects being unified, balanced, serene, tragic, delicate, vivid, moving, trite, and garish [154]. Art can and usually does have aesthetic properties. However, the artist's intentions, the artwork's place in history, its political and social dimensions are also relevant to art. These aspects of art fall outside of what we might regard as "aesthetic." In the discussion that follows, we will examine art while recognizing that aesthetics, often but not always, is a fellow traveller.

An enduring idea is that art depicts the world. Art is imitation. Plato and other Greek scholars developed this idea. In fact, art as imitation made Plato suspicious of art because it distracted us from the real thing. More than two millennia later, in the twentieth century, the art historian Ernst Gombrich characterized the history of Western art as a long process driven to get better at rendering reality [155]. For example, the Renaissance struggle with how to convey three-dimensional scenes from a single perspective on a two-dimensional plane was about imitating the world on canvas. For many people, the ability to render objects accurately defines an excellent artist. People who would be suspicious of a disjointed Picasso painting might be open to his talent when they find out that he was a master draftsman and could draw objects with exquisite fidelity. The fact that he actively chose to draw figures in weird, distorted ways invites the viewer to explore why. Regardless of the desires to render the world accurately, the advent of photography made the goal of painting to imitate the external world less relevant. Soon after the invention and popularity of photography in the nineteenth century, artists were free to experiment with different ways of looking at the world, whether it was with an impressionist, or fauvist, or futurist eye.

Another old answer to the question what is art? is that art is ritualized behavior that brings a community together by solidifying communal values. Before the eighteenth century, much of what we regard as art was done in the service of the church or the state. Art served to elevate us communally. Watching plays by Aeschylus and Sophocles probably brought Greek audiences together. Beautiful hymns and chants in churches and temples bring people together. As a child in India, I remember watching "Ram Lila" street performances. Traveling actors depicted episodes from the Indian epic *Ramayana*. My friends and I delighted in watching these actors depict scenes from the epic as we sat on dirt streets on warm summer nights. My current city has a "One Book One Philadelphia" program, which encourages us all to read the same book. The idea that we all participate in a ritual of appreciating art that brings us together is wonderful. But this feel-good idea of art bangs its head against contemporary art. To what community do cans of shit bind us? Maybe modern art brings together a small cultural elite. For most people, it can confuse, shock, and even alienate.

Alexander Baumgarten's 1750 book, *Aesthetica*, marks the beginning of modern aesthetics. He linked the term *aesthetics*, then associated with sensations, to the appreciation of beauty producing what he called "sensitive cognition." He focused on our reactions to beautiful natural landscapes rather than on art per se. Like Baumgarten, the philosopher Frances Hutcheson thought we have a special sense that is receptive to beauty and harmony and proportionality [156]. This aesthetic sense, Hutcheson claimed, evoked a feeling of pleasure in the beholder. The philosopher and political theorist Edmund Burke also emphasized and expanded on the way beauty evokes feelings. He distinguished between the beautiful and the sublime [157]. Beautiful objects are tied to pleasure. By contrast, sublime objects overwhelm us, produce awe, and force us to face our own insignificance. For Burke, the sublime is tied to pain. The idea that beauty evoked emotions found a parallel in nineteenth-century discussions of art. The Romanticists thought that emotional expression was the essence of art. Art was meant to excite and arouse us. In his 1896 book, *What is Art,* Leo Tolstoy hammered the importance of emotions experienced by the viewer when facing art.

People do not always agree on which art is beautiful. Do we resign ourselves to thinking that beauty in art is all subjective? To avoid this conclusion, the philosopher David Hume, in the eighteenth century, developed the notion of "taste" [158]. He viewed beauty as a pleasure that involved a value judgment. These value judgments were an expression of taste rather than the result of logical analysis. Taste, he thought, might start out being automatic and spontaneous, but it could be nurtured with careful observation and education. People develop a special sensitivity to aesthetic qualities in the world. As taste matures, these sensitive people are best at deciding what is art, or at least what is good art. Hume recognized that education and culture profoundly influence our experience of art.

When it comes to beauty and aesthetic experiences, Immanuel Kant is a giant whose ideas for better and worse continue to influence scientific aesthetics [159]. He thought beauty was an innate and universal concept and that judgments of beauty were grounded in features of the object itself. The judgment of beauty was placed within the realm of reason rather than simply a reflexive emotional reaction to objects. He thought that features of beautiful objects interact with our perception, intellect, and imagination. Kant's general view is compatible with how scientists typically approach art, which is to try to figure out the dynamics of those interactions. Important for our purpose, Kant emphasized the notion of disinterested interest, which could be recast as liking without wanting. For Kant, aesthetic experiences were contemplations of objects that allow our imagination to play freely. This free play happens without the impulse to own or consume the object. In the early twentieth century, Edward Bullough picked up this idea and argued that the aesthetic attitude involves adopting a psychological distance toward the art object [160]. This distance removes practical consideration of objects and opens people to new and deeper experiences that are personally emotional and the root of any aesthetic experience.

Other more recent theorists have different takes on aesthetic experiences. As art became more abstract, art theorizing in the early twentieth century became more formal. Clive Bell introduced the idea of "significant form," which refers to particular combinations of lines and colors that excite aesthetic emotions [161]. For Bell, aesthetic emotional responses were different from other emotional responses. The photograph of a lover might engender desire; the statue of a hero might arouse admiration; the painting of a saint might evoke faith. For Bell, these emotional responses are perfectly normal, but they are not *aesthetic* responses. The aesthetic response is to the forms and relations of forms themselves, not to meanings and memories evoked by the image. Significant form is the reason abstract art can be appealing even if it does not seem to signify any object or meaning in particular.

By the twentieth century it became clear that art could be divorced from beauty and pleasure. The fractures of cubism, the mania of abstract expressionism, the randomness of Dadaism, even when individual pieces might be beautiful, cast a supercilious eye at a naïve infatuation with beauty. So, if traditional views of art as imitation, emotional expression, creation of communal cohesion, and depicting beauty fail to gather all art within its fold, how should we think about what art is? What do contemporary philosophers say? A useful review of their thoughts on defining art can be found in Noël Carroll's edited book, *Theories of Art Today,* and Steven Davies' book, *The Philosophy of Art.* Let me present a few highlights to convey the flavor of these discussions.

One view, popular in the mid-twentieth century, was that art could not be defined. Philosophers call this view anti-essentialist. That means there is no essential ingredient that allows us to say that an object is art when it contains this particular ingredient. Morris Weitz makes two arguments for this position. First, art is inherently revolutionary. Art's rebellious nature undermines itself. Any attempt to confine art with a definition is bound to fail. Define art and some artist somewhere flouts those definitions. Second, art cannot be defined with necessary and sufficient conditions. Rather, art is a collection of objects with a family resemblance. When confronted with a new object, we judge whether or not it is art based on how well it resembles objects we already accept as art. The family resemblance argument is tricky because it doesn't tell us which features are important for the resemblance. A recent version of the family resemblance idea comes from Berys Gaut, who argues that art is a cluster concept. Artworks contain a list of possible attributes. When we find a subset of the list in an object, we call it art. We are still left with the problem of having to pick some properties as more relevant than others to the cluster.

Another argument for art is that maybe we really do know what art is, even if we can't define it. In a 1964 case about the right of free speech, U.S. Supreme Court Justice Potter Stewart famously said about defining pornography, "perhaps I could never succeed in intelligibly doing so. But I know it when I see it." Perhaps art is like Stewart's pornography. Even though we can't define it, we know it when we see it. The philosopher William Kennick used this intuition about art in describing a scenario with a burning warehouse that stored art as well as other objects. He suggested that ordinary people would know which pieces to grab as art and which not to, even if they did not understand a philosopher's definitions of art. Given some contemporary art, that prediction is not so certain. How many of us would desperately grab a urinal as we sputter our way to open air?

To get around the definitional problems of understanding art raised by anti-essentialist philosophers, some philosophers suggest that we can have a coherent concept of art by understanding its relational properties. These relational properties may be how art functions in our lives, or how it is situated in history and culture. The functionalist position suggests that art is understood through our interactions with this special class of objects and how they function in our lives. These objects produce aesthetic experiences that are different from our reactions to other objects. Scientists who study aesthetics are sympathetic to this way of thinking about art. When we construct experiments to probe psychological mechanisms or identify neural markers, we are trying to understand interactions between art and the person. In some ways, functionalist arguments shift the focus from defining the underlying properties of the art object to what is crucial about the encounter with an object that we regard as an aesthetic experience. If that object is an artifact (as opposed to a natural object, like a flower), then we call it art. The functionalist position, as the name suggests, also emphasizes the function that these objects play in our lives. Evolutionary psychologists are sympathetic to this view, since they ask if art is so pervasively present in human history, what function does it serve in promoting human survival.

Another contemporary view emphasizes art as understood by its relation to culture and history. This approach to art falls outside the realm of aesthetics since it is not concerned about the sensory properties of the artwork and the emotions evoked in the viewer. Arthur Danto suggests that the status of an artwork depends on its place within an ongoing narrative and theoretical discussion about art. Similarly, Noël Carroll and Jerold Levinson emphasize the status of art as being fundamentally connected to its precedents. George Dickie has stressed the role of social and institutional practices that combine to designate an object as art. Just looking at something does not tell us that it is art. When Duchamp placed a urinal on a pedestal in 1917 and titled it *Fountain*, why did anyone consider it art? When Robert Rauschenberg painted his bed in 1955, why did anyone consider it art? Arthur Danto used the example of Andy Warhol's 1964 *Brillo Box* to make the point that physical properties of objects are irrelevant to their status as art. Warhol's box is almost indistinguishable from the commercial Brillo boxes mass-produced by Proctor and Gamble. Yet, one box is deified as art, and the others are just containers for a product to be bought and sold. Art, it turns out, is a cultural artifact and can only be understood in its historical context and by cultural practices.

With these dizzyingly different views of art, I am reminded of the well-known parable of the blind men and the elephant. The men touched different parts of the elephant and thought they were touching different objects. One touched a leg and thought it was a pillar. Another touched the back and thought it was a wall. A third touched the tusk and thought it was a spear. Perhaps philosophers of art are doing something similar. They are touching different but true aspects of art. If we are patient, with enough touches, we will gather a complete view of the art elephant.

The problem with this parable is that it only works if we are not blind and can already see the whole elephant. If we are also blind, as we might be with art, the parable is easily turned around. Perhaps one blind man is actually touching a pillar, the other a wall, and the third a spear. In their communal good-natured way, each is imagining touching a different part of the same elephant, when there is in fact no elephant in the room!

As we continue with our exploration of art, tuck this question in the back of your brain. Are we touching different parts of an elephant, or are we imagining an elephant where there is none?

Chapter 2

Art: Biology and Culture

Art is everywhere. In my neighborhood in Philadelphia, I cannot walk a block without seeing something that could be art. Down the street is a mural made of mirror and ceramic fragments. On the sidewalk is a tilted post painted to look like the leaning tower of Pisa. The corner coffee shop rotates paintings and photographs made by local residents. A few blocks away people drink wine and eat cheese every "First Friday" as they swarm galleries. Within walking distance is a serious temple of high art, the Philadelphia Museum of Art. The Barnes Foundation, with its fabulous collection of early twentieth-century art, just moved from the suburbs to join the Rodin Museum and the Philadelphia Museum of Art on the Benjamin Franklin Parkway. The Parkway is anchored by City Hall at one end and the art museum at the other and bisected by the fountain of Logan Square. All three locations—City Hall, Logan Square, and the foyer of the museum—house sculptures by successive generations of Calders.

Philadelphia may be a particularly good city for art. But let me repeat, art is everywhere! Parents hang their children's crayon drawings on refrigerators. People decorate their homes and offices with treasures and trinkets. Every abandoned urban nook and cranny seems tagged with art. The same abundance of art is present wherever you look across the world. If people are gathered, something decorative pops up. This artistic exuberance shows up in clothing, jewelry, wall hangings, pots, pans, murals, and masks. There is art that pleases other senses. We sing and hum and rhyme and rap and beat and tap. We sniff perfume and sip wine. We arrange flowers. We design gardens. We delight in dance. We lose ourselves in movies. We find ourselves in literature. Art is around us, it is in us, it connects us, and it consumes us.

Not only is art everywhere, it would seem that we have always had art. Before the Renaissance there was Persian and Byzantine Art. Before that, there was Roman and Greek art. Further east, there was Indian, Chinese, Korean, and Japanese art. Earlier still was Mayan, Olmec, Egyptian, Sumerian, Babylonian, and Assyrian art. People painted in caves over 30,000 years ago. They decorated with pigments and collected and wore beads 80,000 years ago. They fashioned figurines in North Africa over 300,000 years ago.

If art is everywhere and it has always been with us and we so enjoy it, something about it must be vital to our being, the way that food and sex are vital to our being. Claiming art's vitality is but a short step away from saying it serves an important adaptive purpose. Surely, we must have an instinct for art that is hard-wired in the brain.

The view of art as a biological imperative clashes with the view of art as a cultural artifact. The cultural-artifact view sees art as fashioned locally and our deification of art as a recent invention. According to this view, eighteenth-century European philosophers and their intellectual descendants invented art as we know it. Philosophers like Hume and Kant laid the theoretical groundwork for us to think of art as special objects. If people of taste and education or institutions designate art, then art has to be a product of culture. We might regard urinals on pedestals as high art. Artifacts from other times, whether to glorify gods, banish spirits, or aggrandize the powerful, are simply not art because the people making these artifacts did not have our conception of art.

The idea that art is a cultural artifact is supported by the way we have historically categorized art. In earlier times, distinguishing art from craft and artist from artisan was not particularly meaningful. Music was part of mathematics and astronomy; poetry was part of rhetoric. Artists, if they were recognized as such, served the rich and the religious. Only in late eighteenth-century Europe did galleries, public exhibitions, salons, and art academies emerge. Democratizing art institutions, beyond Church and State, coincided with a growing middle class. The meaning of art changed along with its supporting cultural institutions as it went public [162].

Strong arguments that contrast art as a biological imperative with art as a cultural creation don't make much sense to me. The clash caricatures both biology and culture. In this formulation, the biological view is static, unchanging, and without room for flexibility. It posits a universal disposition to make and appreciate art that is common to all human beings. The cultural view is free and malleable. It posits that we can only understand art by appreciating its history and culture. Neither view satisfies. Our brains are plastic. If they were not, we could not learn or change or grow. When you learn to ride a bike or read a book or sing an aria or play the piano or dance the fandango, your brain changes. This brain change is the physical underpinning of what it means to learn. Whatever makes one person's thinking similar to or different from another's happens in their brain and not in their heart or their liver, not in the air or the ether. At the other extreme, it makes no sense to think that culture, with all its richness and diversity, is disconnected from the human brain. To the extent that culture emerges from collectives of people, and these people have brains, culture emerges from a collection of brains. Undoubtedly, culture and the brain influence each other in complicated ways. Given these multidirectional influences, it makes sense to stop trying to explain art as one or the other; rather, it makes sense to try to understand the way art can be understood both biologically and culturally. Some questions will be better answered by biological methods and others through cultural analyses.

From the vantage point of neuroscience, we can ask which systems in the brain seem preprogrammed and follow an expected trajectory, and which systems are especially plastic and subject to change because of environmental conditions. To illustrate these points about brain plasticity, consider vision and language. When we are children, we all develop depth perception unless we have ocular or neurological disease. The information from our retinae goes to our occipital cortices and, because of a particular design, information from both eyes is combined in specially organized neurons to give us depth perception. We cannot avoid learning to see depth as long as our eyes are open and aligned and we interact with our spatial environment. This preprogrammed learning, triggered by environmental exposures, is hard-wired.

Something similar happens with language. As long as we are exposed to people talking and we do not have neurological disease, we learn language. This learning is also pretty much hard-wired, but with a difference. The languages that different people learn sound nothing alike. While the Czech and Chinese see depth in the same way, they are not likely to understand each other's speech. A major mission of academic linguistics has been to uncover the "universal grammar" that underlies different languages. Language, like depth perception, is acquired in a preprogrammed way that is driven by biology. But, unlike depth perception, language expresses itself with tremendous surface differences that are shaped locally. Beneath these surface differences lies a deeper unity in the structures of all languages. Seeing this deeper unity requires careful excavation.

What about reading and writing? Here we see something different from spoken language. We must be taught to read and write. No amount of staring at little squiggles on a page makes us literate. Reading and writing are learned behaviors that have not always existed. The earliest known examples of writing date back to between 3000 and 4000 BCE. These early examples have been found in different parts of the world, including Mesopotamia, Mesoamerica, China, the Indus Valley, and Egypt [163]. Scholars question whether people in these areas developed writing independently or if writing diffused by cultural contact. Regardless of the answer to that question, writing systems followed centuries of "proto-writing" in the form of pictograms. An interesting fact about writing is that it does not emerge necessarily when groups of people organize themselves in complex social structures. The impressive kingdoms in Hawaii and Tonga and Sub-Saharan West Africa and the largest Native American groups along the Mississippi River did not have writing. The vast Incan empire reigned without writing from the thirteenth to the sixteenth century. Thus, writing is not an obligatory product of our evolved brain. Writing is better considered a plug-in, a cultural tool grafted onto other built-in properties of the brain.

Even though writing is a cultural tool, it has its own place in the brain. Brain damage can give people very selective deficits in reading and writing. In 1892, the French neurologist Joseph Jules Dejerine described a condition called pure alexia [164]. People with this disorder cannot read even though that they can write. Recent fMRI studies have shown that part of the left occipitotemporal region harbors an area that organizes words [165]. This area is now referred to as the visual word form area. Here is a mental ability that is not preprogrammed to appear in our brains, like depth perception or the acquisition of oral language. Yet it is "hard-wired" in the brain. Reading plays a profound role in many of our lives.

It allows us to enjoy Salman Rushdie, Haruki Murakami, Eudora Welty, Joseph Conrad, Gabriel Garcia Marquez, Ben Okri, Vladimir Nabokov, and Toni Morrison. It is a cultural tool that we learn, a tool that gives us great advantages and pleasures, and a tool that now seems indispensable.

To return to our examination of art in the context of biology and culture, should we think of appreciating art as analogous to perceiving depth, to understanding oral language, or to learning to read? Is appreciating art hard-wired as an instinct? Does it have an underlying universal grammar despite its surface differences? Is it a cultural artifact, perhaps etched in our brain, profoundly important in our lives, but not something that contributed to the survival of our ancestors?

Keep this tension, between art as a biological imperative and art as a cultural artifact, in mind as we proceed. Somehow, we will have to reconcile these two views. But before attempting that reconciliation we will look at what biology or, more specifically, neuroscience has to say about art. From there we will meander through questions that pose problems for neuroscience—questions better tackled by history, sociology, and anthropology—before delving into the evolutionary basis for art.

Chapter 3

Descriptive Science of the Arts

What do you think of when you hear the word *bird*? Many think robin. A robin is a good bird. It is a better bird than an ostrich. Ostriches seem too big to be birds and they can't fly. What about a platypus? A platypus just doesn't seem like a bird, even though it has a beak and lays eggs. In the 1970s, the psychologist Eleanor Rosch and her colleagues developed these intuitions about what makes something a good example of a category into what is called "prototype theory" [54]. According to this theory, many categories do not have clear and distinct boundaries. Rather, we understand a category because some members fit a category better than others. When neuroscientists and psychologists study art, they typically focus on prototypes, rather than on artwork that seems marginal. One consequence of this reliance on art prototypes is that we scientists rarely consider contemporary art or converse with artists and art critics in designing our studies.

The early wave of writings by neuroscientists on aesthetics is what I call descriptive neuroaesthetics. This kind of scholarship uncovers parallels between what artists are doing and the way our brains process information. The general idea is that artists, by dint of their special talents, make explicit the mysteries of the way we see the world. These artists sometimes anticipate discoveries later made by neuroscientists investigating how the brain processes vision. In what follows, I will give examples of the parallel worlds inhabited by artists and neuroscientists.

The vision neuroscientist Semir Zeki, who famously coined the term *neuroaesthetics*, pointed out that artists at the turn of the twentieth century honed in on different visual attributes just in the way that neuroscientists have since done. During World War I, scientists first realized that our brain dissects the visual world into different attributes. They encountered many veterans who returned from the battlefield with shrapnel injuries to their brains that picked off different parts of their vision with extraordinary precision. One soldier might have lost his ability to see color, another to see form, and yet another to see movement. The British neurologist Gordon Holmes carefully studied the pattern of their visual problems and worked out the basic organization of our visual system [166]. Information from our eyes goes to the back of the brain, in the occipital cortex, and then is fragmented into distinct attributes (e.g., color, luminance, shape, motion, location) that are processed in different brain areas. Because these attributes make up our visual world, it is no surprise that artists explicitly explored these attributes even before Holmes worked out the fundamental elements of our visual brain.

Zeki realized that rather than use visual attributes to create a realistic rendition of the world, artists were exploring the properties of our visual system itself [167]. Fauvists like Henri Matisse and André Derain, and members of the Blaue Reiter group like Wassily Kandinsky and Franz Marc recognized that one does not need color to define form. Instead, they used color to express emotions. By contrast, the cubists Pablo Picasso, George Braque, and Juan Gris were concerned with form and tried to show that we can represent visual forms of objects without being restricted to one point of view. Duchamp tried to capture motion in his *Nude Descending a Staircase.* The futurists, as announced in the *Futurist Manifesto* by Filippo Tommaso Marinetti, also focused on motion in the context of speed and technology and the dizzying pace of the early twentieth century. Calder was most successful in isolating motion with his mobiles. He reduced shape and color to their simplest forms; the strength of his art lies in the movement of different parts in relation to each other [168].

Artists often focus on attributes that our visual mind finds salient, rather than the way objects truly are in the physical world. The vision scientist Patrick Cavanagh points out that artists often violate the principles of light and shadow and color of the physical world [169]. Typically, we don't notice these violations because they do not clash with the way we represent objects in our minds. For example, artists accurately depict shadows as having less luminance than the object casting the shadow, but they often do not depict the forms and contours of shadows accurately. Our experience of the shape of shadows is too fleeting and changeable to provide reliable information about objects in the world. Our brains were never tuned to notice these shapes. Similarly, transparency is not depicted accurately in art. Images from ancient Egypt show transparency with simple straight crossings rather than with the bend one would expect with optical refraction. For example, if we look at a pencil in water at an angle, we see the line of the pencil bend even though we know that the pencil remains straight. Since we know that objects don't actually bend when they enter water, we don't notice paintings that depict refractions with straight rather than bent lines. Cavanagh says that artists develop and use shortcuts that communicate information about objects without slavishly adhering to the properties of the way objects actually appear when we look at them carefully. The artists' strategies work precisely because our brains evolved to notice only a subset of "true" visual features.

The neurologist Vilayanur Ramachandran offers similar speculations about how artists, at least implicitly, understand the way the brain processes visual information. Ramachandran proposes several principles as art "universals" [67]. We encountered the most important one, the peak shift principle, when talking about our responses to beauty. This principle is that we have an established response to a specific stimulus, and we respond even more vigorously to exaggerated versions of that stimulus. Ramachandran uses the peak shift principle to describe what is going on in bronze sculptures of the twelfth-century Chola dynasty in Southern India. The goddess Parvati is depicted with exaggerated sexually dimorphic features. She has large breasts and hips and an exceedingly narrow waist. He claims that this form depicts the epitome of female sensuality, grace, poise, and dignity and gains its power as an art form by taking advantage of the peak shift principle. He goes on to speculate that our response to abstract art is a peak shift from a basic response to some original stimulus, even when we don't know or remember the original stimulus.

The philosopher William Seeley suggests that artists use techniques to guide our attention across an artwork [170]. He refers to artworks as "attention engines." In a similar vein, the vision neuroscientist Margaret Livingstone and her colleagues have explored how artists produce specific effects to exploit the way our brains are designed to process visual information. These effects could be the shimmering quality of impressionist painters, and even the enigma of Mona Lisa's smile [171]. As I mentioned before, the visual brain dissects vision into the elementary attributes of form, color, luminance, motion, and spatial location. A basic tenet of neuroscience is that these attributes are sequestered in two interacting streams [172]. Form and color are processed in one stream and tell us the "what" of an object. Luminance, motion, and location are processed in the other and tell us the "where" of an object. Livingstone suggests that the shimmering quality of water or the sun's glow on the horizon seen in some impressionist paintings (e.g., the sun and surrounding clouds in Monet's *Impression Sunrise*) is produced when objects are painted with the same luminance but with different colors. The "what" stream sees objects with the same luminance, but the "where" stream cannot. So these isoluminant objects in paintings appear to shimmer because our brains can't fix their location precisely.

Livingstone also explains why Mona Lisa's smile is enigmatic. Our visual system is sensitive to different visual frequencies. At the center of our visual field where we direct our gaze, we see details clearly. Details are conveyed in high-frequency information. By contrast, our peripheral vision is sensitive to broad changes in bright and dark, or is sensitive to low frequencies. So high-frequency vision is like seeing the trees, and low-frequency is like seeing the forest. If we take the *Mona Lisa* and filter the image to keep just the high frequencies or just the low ones, we discover something interesting. The smile is only obvious in the image with low-frequency information and is not seen in the image that has only high-frequency information. Livingstone suggests that when we notice the mouth in our peripheral vision because we are looking at the background of the painting, we see her smiling. When we look directly at her mouth, the smile disappears. It is as if we sensed someone smiling at us only when we are not looking at them. This ambiguity makes her smile seem so enigmatic. Is she really smiling?

In some quarters, the parallel worlds of artists and neuroscientists have made it fashionable to claim that artists, whether Proust or Cezanne, are really neuroscientists. This is a clever idea that grabs our attention but should not be taken literally. Artists are certainly expert analyzers of their world. Of course, some of their ideas and techniques are compatible with what we have learned about the brain. How could they not be? Any field that has developed sophisticated ways of creating objects that fit human needs and desires would have to be compatible with how our brain works. Architects design complex ways to structure our space and guide our movement. Some of their principles of design undoubtedly fit with facts of the brain. Chefs create amazing combinations of tastes and smells that titillate and delight as well as nourish us. These combinations undoubtedly have neural counterparts. Actors are experts at creating expressions, gestures, and conversations that seduce us into suspending disbelief. Their skills at communicating tap into complex aspects of the neurobiology of how we know and understand each other. In the earlier chapter on money, we saw that casino operators really understand the way our brains respond to rewards. Should we proclaim that Frank Gehry, I. M. Pei, Rachel Ray, Emeril LaGasse, Morgan Freeman, Helen Mirren, and even Donald Trump are all really neuroscientists? The claim that artists are neuroscientists does not do justice to the process and rigors of making art or of doing science. To say that an artist is a neuroscientist is like saying that a platypus is a bird because they both have similar protuberances in front of their faces.

To date, descriptive neuroaesthetics has focused on parallels between visual properties of art and the nervous system's organization of the visual world. However, artworks depict more than visual properties. They also convey emotions. Expressionist theories of art emphasize this function. Art can communicate subtle emotions in a way that is cumbersome, if not impossible, to do with words. Art clarifies emotions and, when successful, distills them into their essence. We all, at least implicitly, know that art communicates emotions, just from the words we use. Noël Carroll points out that many of the terms we use to describe people's moods are used to describe art [173]. For example, we can describe people as melancholic, ebullient, placid, joyful, vibrant, gloomy, morbid, or humorous. We use these very same terms to describe art.

To my knowledge, neuroscientists have not seriously considered the expressive qualities of visual art. Presumably, the same logic that says that visual properties of art reflect visual properties of the brain also applies to emotional properties of art. Expressive art might provide pointers to the organization of the emotional brain that neuroscientists have yet to delineate, and neuroscientists might have something to say about how art triggers emotions in our brains.

A neuroscientist might ask several questions about art and emotions. How do lines and colors splattered on a piece of canvas, paper, or wood convey emotions? Is there something special about aesthetic emotions that makes them distinct from other emotions? What exactly is communicating the emotion? Do we as viewers simply recognize emotions in an artwork, or do we also feel them?

Let me offer some preliminary conjectures on how expressive art might relate to emotions. The most straightforward situation would be when paintings depict people's emotions, whether it is the melancholy of a Rembrandt self-portrait or the terror of Munch's *The Scream*. Most of us are experts at reading emotions in faces. The psychologist Paul Ekman has shown that people express and recognize basic emotions, such as anger, disgust, fear, happiness, sadness, and surprise, similarly across cultures [174]. When such emotions are expressed in portraits, the same neural machinery engaged in recognizing these emotions in real life gets engaged during the aesthetic encounter. We are also expert at evaluating landscapes. We know when an area looks inviting and when it looks dangerous. Much of this general expertise was built into the brains of our wandering Pleistocene ancestors. Painted landscapes that similarly convey comfort or foreboding would engage this same neural machinery. Less clear is how abstract images convey emotions. Why do we sense a manic energy in Pollock and settle into calm with Rothko's paintings? Why do we associate red with anger and blue with sadness? Why do rounded shapes please us and jagged angles make us wary? The principles of mapping elemental visual properties such as shape, color, movement, and location onto their emotional tones have yet to be worked out.

We experience emotions at different levels. At the highest level is the interaction of emotions and our cognitive systems. The way we view a situation influences the emotional reaction we might have to a situation. Psychologists have developed this idea into what is called the "appraisal theory of emotions" [175]. The idea is that we interpret objects and events in the world in light of our goals and desires. Subjective states influence the emotions triggered by these objects and events. That is why the same object (or painting) can produce anger in one person, curiosity in another, and amusement in a third. The neural underpinnings of the interactions of people's subjective states, their goals and motivations, with the emotions triggered by art remain to be studied. Art objects trigger another level of emotions, moods, in the viewer. Evoked moods do not necessarily latch onto the art objects directly. For example, listening to a piece of music might make us feel sad or ebullient. Maybe it amplifies stirrings of emotions we are already feeling. These feelings might persist well beyond the specific period during which we are listening to the piece or gazing at a work of art. The biology of how such art triggers neural activity within parts of our limbic system and releases a cascade of hormones into our bodies remains to be worked out.

A level even more basic than moods is reflexive emotions. Certain images produce immediate reactions, such surprise or laughter. Some immediate reactions, like disgust, undoubtedly play on adaptive universal responses. Others are probably biased by our personal experiences. These emotional reflexes seem unmediated by thought. Typically, they produce very quick changes in pupil size, heart rate, and skin conductance—all hallmarks of our autonomic nervous system kicking into gear because of deeply embedded flight and fight responses common to all mobile organisms.

The descriptive form of neuroaesthetics brings empirical scientific knowledge to the discussion of artwork in interesting and often insightful ways. It offers a first draft of the blueprint of how art relates to the brain. Descriptive neuroaesthetics also calls for some caution. It can seduce us into thinking that our knowledge rests on firmer ground than it actually does. The danger is that we start to think of the conjectures of descriptive neuroaesthetics as foregone conclusions. Regardless of how clever or plausible, a conjecture remains an untested hypothesis. To solidify conjectures we need experiments. We need to test predictions generated by conjectures of descriptive neuroaesthetics. In other words, we need experimental neuroaesthetics.

Chapter 4

Experimental Science of the Arts

Gustav Fechner was a major pioneer of experimental psychology and empirical aesthetics. In 1860, he published *Elemente der Psychophysik (Elements of Psychophysics)*, a treatise on how we quantify sensations. Fechner discovered that our psychological experiences of sensations are lawfully related to physical properties (such as brightness or loudness) of objects in the world. In 1876, he published *Vorschule der Aesthetik (Primer of Aesthetics)*, a treatise that extended his psychophysical methods to aesthetics.

Fechner's experiments mark the beginning of scientific aesthetics. His approach to aesthetics was "bottom-up." Bottom-up means that he investigated how simple visual features, such as size, shape, color, and proportion, affect people's preferences. For example, he conducted early studies on the golden ratio (the proportions we saw in the chapter on beauty in math) to find out which kinds of rectangles please people. His psychophysical studies of the contribution of simple visual features to aesthetics spawned countless experiments over the subsequent century and a half. His methodological innovation was to take many examples of simple stimuli and average the reactions of many people. After Fechner, investigators could use statistics to test hypotheses rather than rely on their own insights or the subjective sensibilities of one or two people.

Is reducing our visual world to its elements the best way to study perception or aesthetics? The Gestalt psychologists of the first half of the twentieth century did not think so [176]. Three German psychologists, Max Wertheimer, Kurt Kofka, and Wolfgang Köhler, advocated a different approach. They thought that examining how simple visual features impinged on our minds made the whole process seem too passive and was the wrong way to think about perception. Instead, they postulated that we see the world holistically. Our minds actively organize visual elements into more complex chunks, thus scientists should be studying these organized chunks. They described several chunking principles that have names like proximity, continuation, similarity, and closure. The specifics of these principles are not important for our discussion, other than to recognize that without some organization, the world would be a blooming, buzzing confusion of inchoate visual elements. The Gestalt approach to perception as applied to art reached its peak in the mid-twentieth century with studies conducted by the psychologist Rudolph Arnheim [177]. Arnheim emphasized formal principles such as balance, symmetry, composition, and dynamic visual forces as critical ingredients in the appreciation of art.

The next major trend in empirical aesthetics was a move from perception to the role of attention and emotion. This move in the mid-twentieth century also brought the empirical enterprise closer to neuroscience. Daniel Berlyne emphasized the role of arousal and motivational factors in our experience of viewing art [178]. He thought that properties such as novelty, surprise, complexity, and ambiguity in art, properties not considered by the psychophysical or Gestalt scientists, were important. For example, he thought there was an optimal level of complexity in objects that people find appealing. Objects less complex than this optimal level were boring, and objects more complex were chaotic and overwhelming. For Berlyne, these optimal configurations create a state of arousal in the viewer that drives emotional responses in aesthetic experiences. His work linked perceptual and cognitive aspects of aesthetics and underscored its links to neurophysiology.

The approach in empirical aesthetics that asks whether we are reacting to properties of visual images has been revived by scientists who use modern image statistics. This research suggests that artworks contain quantifiable parameters that we find attractive even if we are not explicitly aware of those parameters. For example, fractal dimensions refer to the way patterns repeat themselves at different scales. Fractals are found in irregular but patterned natural shapes, such as branching trees and coastlines. Fractal dimensions occur between 0 and 3. One-dimensional fractals rank between 0.1 and 0.9, two-dimensional fractals between 1.1 and 1.9, and three-dimensional fractals between 2.1 and 2.9. Most natural objects shown in flat images like photographs or paintings have fractal dimensions between 1.2 and 1.6.

The physicist Richard Taylor drew attention to fractal dimensions in art by examining Jackson Pollock's drip paintings [179]. He and his colleagues found that Pollock's early paintings had fractal dimensions around 1.45, the dimensions found for many coastlines. Over time, as Pollock's paintings became richer and more complex, the fractal dimensions climbed as high as 1.72. After making these observations about Pollock's paintings, Taylor found that people preferred artificial images with fractal dimensions between 1.3 and 1.5. Images with these dimensions are neither too regular nor too random.

Taylor's claims became controversial after he used his method to authenticate new paintings purportedly painted by Pollock [180]. He did so at the request of the Pollock-Krasner Foundation and concluded that the new paintings were probably not authentic. However, his conclusions were challenged by a physics doctoral student, Katherine Jones-Smith, and physicist Harsh Mathur [181]. They asserted that Taylor's methods could not, in principle, adequately determine a paintings' fractal dimension. They also showed that a simple line drawing made in Photoshop had the fractal dimension that Taylor thought was characteristic of Pollock's paintings. To my knowledge, the dispute has not been resolved. I am not enough of a mathematician to have an opinion on who is right. However, the dispute played out in *Nature*, one of the most prestigious scientific journals.

Taylor raised the possibility that we respond to hidden mathematical regularities in artworks. Christopher Redies [182] in Germany and Daniel Graham and David Field [183] in the United States independently confirmed and advanced this basic point. These scientists found that visual art and natural scenes share statistical properties, including that they are typically "scale invariant." This property means they contain the same kind of information, whether one zooms in or out of the image, such as when looking at a mountain or zooming into a rock on a mountainside. This is a property of the entire image and not just of specific details. These investigators identified scale-invariant properties of visual art by using the Fourier power spectra of images. Fourier spectra describe the range of spatial frequencies from low (broad swathes) to high (fine details) in any image. For our purposes, without delving into its mathematical details, we need to know that natural scenes have characteristic Fourier spectra. These investigators go beyond Taylor by explicitly linking the statistical properties of art images to the way the nerve cells are thought to process information efficiently [184].

Redies found that the Fourier spectra of many examples of art from the Western hemisphere, ranging from fifteenth-century engravings to twentieth-century abstract art, have values similar to those found in pictures of natural environments. These spectra differ from what one would find in photographs of laboratory and household objects, plants and parts of plants, and scientific illustrations. It looks like artists create art with statistical properties that are not necessarily the same as those found in photographs of objects. Rather, artists create images with Fourier spectral-image statistics seen in complex natural scenes. These image statistics also apply to abstract images. Viewers prefer abstract paintings that match the statistical properties of natural scenes. Redies found that cultural variables, such as technique, century, and country of origin, and subject matter did not change these quantitative parameters in the art. Graham and Fields also examined 124 museum paintings that contained both Western and Eastern artwork and found similar hidden statistical properties [185]. Even more intriguing, Redies looked at the Fourier spectra of photographs of faces and those of painted portraits [186]. He found that painted portraits have statistical properties that are closer to those of natural landscapes than to those of natural faces.

These findings—that we like images with specific quantifiable parameters embedded within them, parameters of which we are typically not aware—are the most recent version of the research program started by Fechner. He has characterized this family of experiments as "outer psychophysics," by which he means there are lawful and quantifiable links between our psychological states and physical properties of the outer world. Outer psychophysics can also be thought of as the study of aesthetic properties of objects, properties that are objective but evoke an aesthetic experience within us. Fechner also recognized the possibility of an "inner" psychophysics, which links psychological states to the physical properties of our brain. In the nineteenth century, scientists simply could not investigate inner psychophysics because the needed technology did not exist. Now, 150 years later, we are ready to enter the inner psychophysics of aesthetics.

Before getting to the inner psychophysics of aesthetics, however, let's review the overall brain organization relevant to encounters with art. As mentioned before, aesthetic experiences have the core triad of sensations, meaning, and emotion, each with different neural underpinnings. Neural responses to sensations themselves would, of course, vary depending on whether the art is received through sight or sound or taste or touch, because each of these sensory systems has its own entrance into the brain. For vision, the focus of this book, processing can be classified as early, intermediate, and late [187]. Early visual processing extracts simple elements from the visual environment, such as color, luminance, shape, motion, and location. Fechner's psychophysics typically focuses on responses to these simple elements. Intermediate visual processing segregates some simple elements and groups others together to form coherent regions. Gestalt psychologists like Arnheim studied this level of visual processing as it related to art, although he did not make explicit reference to the brain. Late visual processing recognizes objects and the meanings and memories and associations triggered by those objects. Along the way from sensations to meaning, emotion and reward systems are activated. Some combination of the activity in neurons that code sensations, meaning, and emotions is the neural manifestation of an aesthetic experience.

Several studies have used artwork to locate aesthetic processes in the brain. Let me first offer a taste of these studies (with no real attempt to be comprehensive) and then draw some conclusions from them.

The neuroscientists Kawabata and Zeki asked people, while their brains were being scanned, to rate abstract, still-life, landscape, or portrait paintings as beautiful, neutral, or ugly. As you might expect, the pattern of activity within the ventral visual cortex varied depending on whether subjects were looking at portraits, landscapes, or still lifes [188]. You would expect this pattern because different regions of this part of the brain are tuned to respond to faces, places, or objects. The orbitofrontal cortex and anterior cingulate (important parts of the reward system, as we saw earlier) were active for beautiful images. This is the same area that is active when we experience pleasures of different kinds. Oshin Vartanian (remember his fondness for Wynona Ryder?) and Vinod Goel also used images of representational and abstract paintings in a similar fMRI study. They found that activity within the occipital cortices and the left anterior cingulate increased the more people liked paintings [189].

What about brain responses to beauty for abstract images? Jacobsen, Schubotz, Höfel, and von Cramon used a different strategy to address this question. Rather than use actual artworks in their experiment, they used geometric shapes designed in the laboratory. Participants judged whether the images were beautiful or whether the images were symmetrical. This strategy ensured that participants looked at the images carefully while making judgments, and that the basis for their judgments differed depending on the condition. The scientists found that aesthetic judgments of beauty, more than symmetry judgments, activated medial and ventral prefrontal cortex, as well as a part toward the back of the brain called the precuneus [190]. These regions are part of the extended reward circuitry.

MRI is not the only method used to study the relationship between mind and brain. Camilo Cela-Conde, Marcos Nadal, and colleagues used a technology called magnetoencephalography. This technology records brain waves when participants do specific tasks and is sensitive to the timing of what happens in the brain, as compared to MRI, which is sensitive to the location of what happens in the brain. These investigators had people look at artworks and photographs and judge whether or not the images were beautiful. More beautiful images evoked a bigger neural response than less beautiful images over the left dorsolateral prefrontal cortex 400–1000 msec after the images were shown [191]. This finding highlights that decision-making parts of our brain distinguish beautiful images very quickly, in much less than a second!

Let's consider these studies in the context of the core triad of aesthetics: sensations, meaning, and emotions. First, sensations. Of course, parts of visual cortex responds to visual art. It is no surprise that a portrait activates the face area and a landscape activates the place area. We do not really know if these visual areas of the brain are also involved when we evaluate art. Are these areas responding to the art's beauty? Are they part of the neural basis for the pleasure that we experience from art we like? The study by Vartanian and Goel suggests that neural activity in these areas does increase when people look at images they find more beautiful. You might recall in the section on facial beauty a study from my lab in which we found that visual areas responded to beauty even when people were engaged in a task that had nothing to do with beauty. Perhaps these visual areas are an extension of our pleasure circuitry when it comes to art and beauty. The neuroscientist Irving Biederman has observed that neurons in higher order visual areas in the ventral occipital cortex have opioid receptors [192]. As we saw in the section on pleasure, opioid receptors in the nucleus accumbens receive important chemical signals for our pleasures. Perhaps these receptors in visual-processing areas also signal our pleasure in visual art.

When we consider emotions, we find that the pleasure evoked by viewing beautiful art activates the orbitofrontal cortex, the anterior insula, the anterior cingulate, and the ventral medial prefrontal cortex. These are the same brain structures that good food, sex, and money engage. However, there is much that we do not know about these pleasures. Some studies find activations in some areas, like the orbitofrontal cortex, and not in others, like the ventral medial prefrontal cortex. What distinguishes the experience of these different patterns of activation evoked by different works of art? We know very little about nuanced emotions that can be evoked by art, such as mixtures of fear and disgust, and wonder and whimsy.

Next, consider the role of meaning in art. A short description of an artwork or even knowing the name of the artist changes our aesthetic experience of looking at a painting [193]. People can judge very quickly whether they like a painting, but it takes longer (10 seconds or more) for the descriptions to produce an initial understanding of the painting. People can also be given information that fits or doesn't fit with what they see. The psychologists Martina Jakesch and Helmut Leder found that such dissonant information has a peculiar effect on the experience of looking at abstract paintings. When given ambiguous information, people found modern abstract artworks more interesting and liked them more [194].

What happens in the brain when meaning is attached to art? Ulrich Kirk, Martin Skov, and their colleagues approach meaning in art by asking whether our expectations affect the experience of what we are viewing [195]. If you recall from our earlier discussion about taste, the enjoyment of Coke or Pepsi is affected by whether a person knows the name of the brand of cola they are drinking. Kirk and his colleagues found a similar context effect when people viewed art. Participants in their fMRI experiment looked at abstract "art-like" stimuli that were labeled as either coming from a museum or generated by a computer. People rated the same images as more attractive when they were labeled as museum pieces than when labeled as computer generated. This preference was reflected in more neural activity within parts of the reward system: the medial orbitofrontal cortex and the ventromedial prefrontal cortex. Thinking that an image was a museum piece also produced more activity in the entorhinal cortex, an area tightly connected with the hippocampus and important for memory. Here, as in the Coke-Pepsi study, we see that meaning in the form of people's expectations influences how they experience visual images. These expectations draw on people's memories and can enhance, and in some circumstances diminish, their visual pleasure.

People can bring knowledge to their visual experience just by being familiar with the art. The psychologist James Cutting found that people prefer impressionist paintings just by being exposed to them [196]. The neuroscientists Weismann and Ishai [197] scanned participants looking at Braque and Picasso cubist paintings. Half of the participants were given 30 minutes of information about Cubism and practiced recognizing objects in such imagery. When looking at cubist paintings, these people had more activity in the intraparietal sulcus and parahippocampal gyrus than did people without similar exposure to the paintings. A short training session had an influence on their perception of these paintings that could be neurally recorded.

Investigating people with and without expertise is another way to find out what happens in the brain when knowledge influences visual experiences. One study recruited architecture students as experts in buildings and compared their responses to those of other students as they looked at pictures of buildings and faces [198]. The architecture experts had more neural activity in the hippocampus when they looked at buildings than when they looked at faces. This neural response suggests that the pictures of buildings activated their memories of buildings. When looking at buildings, they also had more neural activity than non-experts in parts of the reward system: the medial orbitofrontal cortex as well as the anterior cingulate. In these examples, the architecture students' building expertise modulated their pleasure. By contrast, there was more neural activity in the nucleus accumbens for attractive faces and buildings irrespective of viewers' level of expertise. This core pleasure center seems to record our enjoyment of objects regardless of the effects of education and background.

As we survey these results of experimental neuroaesthetics studies, we find that there is no art module in the brain. Rather, our subjective experience when we encounter art is cobbled together from bits and pieces of the brain that are used to do other things. It could have turned out that art has its own special visual circuitry, the way faces and places do. It could have turned out that art evokes a special emotion in its own sequestered brain location. It could have turned out that art has special meaning set apart from our everyday knowledge of the world. But it did not turn out that way. The brain responds to art by using brain structures involved in perceiving everyday objects—structures that encode memories and meaning, and structures that respond to our enjoyment of food and sex.

From this brief tour of the world of scientific aesthetics we see that scientists are beginning to understand how our brains respond to art. It is an exciting time to be working in neuroaesthetics. In the midst of the excitement of this new enterprise, we might step back and ask the following question: Are there limits to the analytical scrutiny of science when it comes to art and aesthetics? I am not talking about the technological limits that Fechner recognized when he envisioned a future inner psychophysics that was not possible in his day. I am talking about limits that in principle are beyond the reach of science. Most experimental studies use well-accepted pieces of art as the focus of their inquiries. How should scientists approach art urinals and Brillo boxes enshrined in museums? One strategy would be to dismiss these kinds of art as marginal. Let's proclaim that these pieces are outliers and weird and set them aside for another day. Let's just deal with those pieces of art that we all agree are art. Alternatively, scientists might try to grapple with recent movements in art and ask: Can scientific aesthetics say anything useful about conceptual art? The next chapter addresses this question.

Chapter 5

Conceptual Art

Consider five conceptual artists and their art. Each of these artists received attention and notoriety. Their kind of art has been called conceptual, post-modern, avant-garde, cutting edge, or emergent. The average person looks at this artwork in bewilderment and asks, why is this art? We can ask, how should scientists think about such art? Should we relegate such art to the margins as distractions, or should we see if science has anything meaningful to say in this context?

A mass-produced crucifix floats in golden amber fluid. The light in the Cibachrome photograph of this crucifix looks ethereal, even reverential. But the golden fluid is the artist's urine. Earlier I mentioned the controversy produced by Andres Serrano's *Piss Christ*. I did not mention that Serrano regards himself as Catholic. His Christianity is personal. In his art, he struggles deeply with his faith and its social institutions. He challenges the notion that the Church has the authority to tell people how to value their bodies and that some bodily fluids are disgusting. *Piss Christ* also refers to Gauguin's *Yellow Christ*, painted a century before. Gauguin rejected European cultural norms as he set out for the distant South Seas. Serrano, mixed Afro-Cuban and Honduran, rejects the polite world of public piety. Crucifixion is an ugly, painful, and horrific way to die. However, Serrano sees the symbol as now sanitized of its horror. He challenges peoples' reverence for religious iconography that often masquerades as reverence for religion itself. His struggle with religion is not obvious from just looking at the picture. Instead, the picture and its title seems to be pissing on good people.

A blue square mat lies on the floor in the corner of a room. Stacks of white paper squares cover each corner of the mat, revealing a blue cross. Visitors are encouraged to take a sheet of paper. One could easily see this artwork by Felix Gonzales-Torres as some minimalist abstraction, more pretentious than interesting and hardly engaging. However, knowing that his partner was dying of AIDS, that he was preoccupied with the transience of life and its limits, and that Blue Cross is the most famous U.S. medical insurance company transforms the experience of looking at the work. The blue cross that emerges between the stacks of paper, and the stacks of paper themselves, becomes a potent symbol of medical insurance and its invasive presence down to the final twitches of our lives. Giving away sheets of paper encourages visitors to engage with the art and to take home a piece of the problem. Gonzales-Torres' art calls the viewer to action, to respond to the problem that inspired his art.

Continuing in a geometric vein, imagine two cubes, each 2 × 2 × 2 feet, weighing 600 pounds. One is dark brown, the other creamy white. Around the edges and corners are curious marks that on closer inspection reveal themselves as tooth marks. The dark cube is *Chocolate Gnaw,* the light cube is *Lard Gnaw.* Janine Antoni is the artist who chewed on these sculptures. Her sculpture draws attention to the pleasure and guilt of consumption, and the relationship between consumption and eating. She undertook this project in three phases. The first phase involved the creation of the cubes. For *Chocolate Gnaw,* she melted chocolate and poured it into 50-pound layers and then allowed them to cool before adding the next layer. For *Lard Gnaw,* she cooled lard with dry ice after filling a mold. The process was monotonous, repetitive, and obsessive. After completing the cubes, Antoni's mouth became her tool in the next phase of this project. She bit the cubes, leaving the marks of her mouth for all to see. She took evolutionary imperatives, the desires for sugar and fat, far beyond any nutritional need. She spit out bits of chocolate and lard and considered the acts of biting and spitting a metaphor for a culture that consumes and discards with impunity. She used these spit-out bits to make new sculptures for the third phase of her project. She melted the chocolate bits to make heart-shaped candy boxes and mixed the lard bits with pigment and beeswax to mold them into bright red lipstick. These symbols of romance and desire were then displayed in a boutique close to the chocolate and lard cubes. Her artwork ruminates on a consumer society that assaults women's sense of themselves and their beauty.

Mireille Suzanne Francette Porte is no stranger to women's struggle with ideas of beauty. Better known as Orlan, she began *The Reincarnation of Saint Orlan* in 1990. She subjected herself to several cosmetic surgical procedures that were broadcast to the Pompidou Center in Paris and the Sandra Gehring Gallery in New York. As part of her reincarnation, she chose the chin of Botticelli's *Venus*, the nose of Jean-Leone Gerome's *Psyche*, the lips of Francois Boucher's *Europa*, the eyes of *Diana* as depicted in sixteenth-century school of Fontainbleau painting, and the forehead of da Vinci's *Mona Lisa*. Her surgeons wore clothes made by fashion designers as they became players in her performance while they operated on her. Orlan described her work as a struggle against "the innate, the inexorable, the programmed, nature, DNA, and God." Life has since caught up with Orlan's art. In 2010, according to the American Plastic Surgery Society, over 13 million cosmetic surgical procedures were performed in the United States. Of these, over 1.5 million were invasive procedures. Nose reshaping and eyelid surgery were among the top five invasive procedures.

In another paean to obsession, a 27-year-old woman followed a man named Henri B over 13 days from the streets of Paris all the way to Venice. She used makeup, wigs, gloves, sunglasses, and hats to disguise herself while she stalked the man. She adopted the trappings of a person pathologically obsessed with another, except for a missing key ingredient. She knew nothing about Henri B. He was a total stranger, purportedly picked at random. The work is *Suite Venitienne* and the artist is Sophie Calle. She says that she has never been jealous enough, or in love enough, to do the things she did to be connected to this stranger. She created the behavioral expressions of an extreme emotion, without actually feeling them. After 13 days, Henri B realized he was being followed and confronted her. Nothing dramatic happened. Calle described this final encounter as a banal ending to a banal story. Her art was to play out empty romances with which we easily delude ourselves. She examined our indulgent fantasies that we easily project onto others.

The philosopher Arthur Danto has called conceptual art *intractably* avant-garde [199]. He uses the word *intractable* to set it apart from earlier artistic movements that also bewildered naive viewers. These earlier movements were initially scorned by critics and later accepted and even adored by many. The prime example is the initial rejection of impressionist paintings by the Salon in Paris. Impressionist paintings are now among the most popular artworks ever produced. Paintings by Van Gogh, Gauguin,

Matisse, Modigliani, and Picasso had similar trajectories. More recently, abstract expressionist art by Pollock and de Kooning, and pop art by Andy Warhol and Roy Lichtenstein have enjoyed enormous popularity.

Danto thinks that most contemporary conceptual art will not follow this trajectory of starting with rejection and then evolving to adoration. Some conceptual art can never become ornaments in a museum or gallery or occupy the homes of Wall Street magnates. He calls this art intractably avant-garde for two reasons. First, it disregards pictorial space. The gilded frame, the rectangle, so familiar to visual art, has been tossed aside. Now, a stack of papers, a hunk of lard, a surgical incision, or furtive surveillance can be art. This art is not designed to please the eye. Neither pictorial aesthetics nor visual literacy helps guide the encounter. These facts about this kind of art will not change over time, and so Danto thinks we will never be quite at ease with it.

Second, Danto thinks this art is intractably avant-garde because it does not develop progressively. The art is fluid, changing, ephemeral, redefining itself, encompassing different endeavors, without a clear trajectory. Practitioners of much of this art resemble activists more than lone geniuses striving in isolation to capture ideals of beauty. Whether such art will survive the power of art galleries, dealers, collectors, auction houses, and blockbuster exhibitions to showcase, sell, and own pictorial spaces remains to be seen. Charles Saatchi, the dealer, curator, marketer, and advertiser all rolled into one, promoted his own exhibit in 2005 called the "Triumph of Painting," as if to reign contemporary art back into a more conservative pictorial space.

Even when contemporary art pleases the eye within a familiar frame, like Serrano's *Piss Christ*, it does not derive its artistic force from beauty. Beauty and pleasure might be naive preoccupations of a bygone era. For many, the works are certainly not about the viewer adopting a stance of disinterested interest that eighteenth-century theorists envisioned. Conceptual art, like that of Gonzales-Torres or Antoni, is meant to actively engage their audience and spur them on to change the world.

The art critic Blake Gopnik also emphasizes the importance of meaning over that of the sensation or emotions artworks evoke [200]. For him, beauty is largely irrelevant. He has almost never used the term in over 500 articles on art that he wrote for the *Washington Post*. Gopnik is less inclined than Danto to think that conceptual art fundamentally differs from earlier art. For Gopnik, art is and has always been about meaning, as understood through its social and historical context. Even when formal properties are prominent, he argues, an artwork's content remains fundamental to its appreciation. For example, Seurat's *Afternoon at the Island of La Grande Jatte* is often described in terms of its pointillist technique, its use of colors, and its novel way of rendering the surface of a picture. However, critics first reacted to the painting's social critique of the banal promenade of maids and clerks and troopers rather than to its stylistic innovations. Picasso's still life with an absinthe spoon, playing cards, and furniture is fully understood as a social commentary on the entry of mass media retailing. For Gopnik, the power of art lies in meaning that tells us something about people and their social worlds. Art historians and critics excavate layers of meaning that make encounters with good art resonant experiences.

When we consider conceptual art, it becomes obvious that these artworks are vehicles for ideas. Reactions to the artwork, whether appreciation or rejection or controversy, are reactions to the ideas embodied in the work. Conceptual art makes explicit the importance of meaning and interpretation in understanding and appreciating art, and points us to their importance lurking below the surface of all art. Without some background information, such as the context in which the work is produced, the intent of the artists, the potential meanings it conveys, the cultural conversation in which it is engaged, we cannot fully understand the artwork.

Does science have anything useful to say about meaning in art? The ultimate reach of science is hard to predict, but to my knowledge there has not been any serious attempt to think about the science of conceptual art. Consider the critical aesthetic triad of sensations, emotions, and meaning. Scientists have typically focused on the connection between sensations and emotions. Art, as long as it slides along this sensation–pleasure groove, is amenable to investigation by empirical methods. Scientists can look for hidden stable regularities of light and line and color and form in artwork that are pleasing and relate them to the kinds of neural tuning for which our brains seem to be designed. We can examine the neural response that accompanies emotions evoked by artwork. Much of neuroaesthetics research thus far has been on pleasure in a fairly simple way. Do you like something? Do you want it? But this simple approach of measuring preferences and pleasure is not a principled limit of what neuroscience can do.

Neuroscience could have something to say about complex admixtures of emotions. Burke and Kant, as we saw earlier, emphasized the idea of the sublime, a kind of beauty that evokes pleasure mixed in with anxiety and fear. *Sublime* was used to describe landscapes in which we experience our limits, our smallness. One recent study found that fear can enhance aesthetic experiences, linking this phenomenon to experiences of the sublime [201]. We are also learning more about the psychology and neuroscience of disgust. Artwork that plays on combinations of pleasure and disgust could also be studied by neuroscience methods.

Unlike sensations and emotions, when it comes to cultural and historical meaning in art, we run into the limits of what neuroscience can offer. Current neuroscientific methods are best at investigating the biology of our minds for properties that are stable and relatively universal. We can apprehend the general meaning of a scene very quickly. Neuroscience does have something to say about how this process works. In the same way that we easily interpret what we see when we look out our window, we easily interpret what we see in representational art. This ability is partly why artistically naive viewers prefer representational paintings over abstract ones: they can latch onto a piece of the meaning of the painting. However, the aspects of meaning in an artwork that changes over time and relies on the interplay of its cultural context, the artist's intentions, and the local biases of the viewer are too slippery to be grasped by neuroscience. The richly textured meaning of individual pieces of art that gives art its power is inherently variable and open to many interpretations and thus closed to neuroscience.

While neuroscience probably cannot investigate complex meanings contained within individual artworks, it can deal with the effects of meaning. Experts and novices engage with artwork differently. These differences can be investigated. For example, experts and novices see paintings differently. Scientists can record their respective gaze patterns to get a sense of which aspects of a painting draw their attention [202]. As we saw before, our background knowledge changes our emotional experience of art and other objects. This effect of background knowledge was evident in the Coke-Pepsi study and in the study in which people preferred patterns that they thought were hanging in museums over those they thought were generated by computers. However, compared to the multiple dimensions of knowledge that can be brought to bear in understanding an artwork, these effects of information and background knowledge on aesthetic encounters are relatively one-dimensional. For example, mathematics played an important role in the intellectual culture of early modern Europe, a role that influenced painting. Barthel Beham, in his 1529 portrait of an unidentified man, shows the sitter working out a math problem. The depicted mathematic algorithm turns out not to make sense [203]. It is not really a coherent math problem. Why would the artist choose to show these numbers and symbols in such a way? These kinds of questions, probing the historical and cultural context in which the painting was made, are not easily addressable by neuroscience. Neuroaesthetic studies could explore the way that meaning influences the encounter with an artwork even if it cannot address the meaning of an individual artwork itself. One could set up experiments to see how learning about the influence of mathematics on visual culture in fifteenth- and sixteenth-century Europe would change the viewer's reaction to the painting. That change could be tracked neurally.

Conceptual art, with its emphasis on meaning shaped by culture, is hard to bring under scientific scrutiny. We saw in the last chapter that neuroaesthetics studies are naturally designed to address the sensation–emotion axis. Conceptual art forces scientists to consider how layers of meaning can be folded into the design of experiments.

Some people might think that conceptual art distracts scientists from getting to the very essence of art. Rather than examine this very new and often confusing art, maybe we should turn to very old art. Examining art at its germination might give us art in a pure form. We might see art's essential ingredients revealed, uncontaminated by excesses of our accelerated culture. Maybe looking backward will give us clues to how neuroaesthetics might move forward.

Chapter 6

The Roots of Art

In 1940, on September 12, four teenage boys and a dog stumbled accidentally into obscure caves in Lascaux, France. Imagine them wandering into these dark spaces, hearing reverberations of their footsteps against the rough ground. Maybe they heard the distant tinkling of water and saw light reflecting off the winding walls and rock formations creating bizarre shadows and patterns. Amazingly, on the walls and the ceiling of these subterranean halls they found an ancient bestiary. Horses and stags and cattle and bison roam in darkness. Four huge bulls are in motion. One bull is 17 feet long. The undulations and protrusions in the walls on which the animals are painted add depth and dynamism to the forms. There is an upside-down horse. Big bears and cats are sequestered in the deepest recesses of the caves. Unlike the vivid animals, a human figure is painted crudely and looks like it is wounded. In addition to the animals, geometric forms are scattered on the walls. Red and black dots, lines, and hatches and geometric figures abound. The art is drawn and painted in black, brown, red, and yellow pigments made of minerals like hematite and oxides of iron and manganese. These paintings were made between 20,000 and 15,000 years ago.

Perhaps, journeying back to the origins of art, before the reach of the Sothebys and Saatchis in our manic world, we might find art in a pristine form. Philosophers point out difficulties in defining art, and critics and historians point out difficulties in interpreting art. The Pleistocene era, the period between 1.8 million years to 10,000 years ago, is when we find traces of early art. Perhaps examining ancient art will give us insight into both art and aesthetic experiences at their very inception.

After walking through Lascaux, Picasso purportedly said, "We have learned nothing." The Picasso anecdote is repeated often, even if it probably did not happen. The anecdote makes a good story. The standard story of early art was also a good story, even if it did not happen. The story is collapsing under the weight of accumulating evidence [204]. The standard story is that modern humans migrated out of Africa into Europe around 40,000 or 50,000 years ago. Eventually, they got to northern Spain and southern France. Along the way, they replaced brutish and backward Neanderthals. While colonizing this land, they erupted with a big bang of creativity. These early humans became artists and created amazing paintings in caves such as those of Lascaux and Chauvet in France and Altamira in Spain. This flowering of human cultural consciousness was shaped in Europe and then exported to the rest of the world.

Several questions raise doubts about this standard story [205]. Was this creative explosion really an abrupt eruption, or did it build up gradually? Can we trace this artistic tradition through history and see how it evolved to influence art of the present day? How far back in time can we go before cave paintings and still find something we might call art? Is art a unique product of modern humans, an outpouring of our modern brains? Does a unifying theme connect ancient artistic efforts?

Identifying the origins of art turns out to be complicated business. As best we can tell, artistic behavior emerged in fits and starts, and pulsed with idiosyncratic patterns. Early humans in Africa, Asia, and Australia used pigments, fashioned bone, made beads, and carved engravings and sculptures much earlier than the standard story posits with its big bang of creativity [206]. Keep in mind, we only know about artworks that have survived over the years. Paleoarcheologists only have artifacts from locations that are safe from the elements and are made of especially durable materials. We have bone and stone remains, but not fabric and fur. Contemporary regional intellectual and academic resources often dictate where we even look for ancient traces of art. Keeping in mind these limits of durability and geography and resources, let's work backward in search of artistic activity that preceded the Lascaux caves.

The remarkable forms and styles of the animals painted in Lascaux exist in other caves in the same general region of Europe. Some are considerably older. Paintings in Chauvet are estimated at being 32,000 years old. To get a sense of these paintings, I recommend viewing Werner Herzog's documentary movie, *Cave of Forgotten Dreams*. The caves at Chauvet are open to a handful of select scientists for a few weeks each spring. Herzog had unprecedented access to the caves, and his movie conveys the eerie power of these ancient beautiful paintings. While later cave paintings have a few stylistic differences from those at Chauvet, what is striking is their similarity. Animals are painted with similar colors in profile. They are not painted with their background environment. Keep in mind that the time between Chauvet and Lascaux is almost as vast as the time between Lascaux and the Louvre. For over 20,000 years, the same animals—bison, stags, aurochs, ibex, horses, and mammoths—were painted in similar poses. The art of Chauvet is not a simple, earlier version of a tradition that found its full expression 15,000 years later in Lascaux. Very early on, the artists of this region developed techniques that included a sense of depth and an amazing ability to capture movement. The artistic conventions of Chauvet continued for over 20,000 years with relatively little change. Without diminishing their incredible artistic beauty and the remarkable innovation that produced these works, it boggles the modern mind that the artists didn't experiment further with different styles, or embed the animals in their environment, or show them from different angles. Nobody painted a landscape! Nobody painted a portrait! Just think of the art of the last century and how every decade dished up something new. By contrast, these ancient artists did variations on the same beautiful work over, and over, and over, for more than 20,000 years. We get the sense that they were highly skilled artisans following prescribed patterns, rather than being the avant-garde of a prolonged age of fervent innovation.

Did art exist before Chauvet and other caves of its time? The period between 300,000 and 50,000 years ago is sometimes called the middle Paleolithic. In this swath of time, we see rudiments of artistic behavior. Besides paintings, we find engravings on rock surfaces and cave walls. While the colorful animals of the European caves are most impressive, geometric forms and various indentations are far more common and are found all over the world. Beads and shells were used as adornments. Some portable pieces were fashioned exquisitely, others slightly modified, and still others may have just been found and kept because of their significance. Let's wander further back in time to look at the range of objects or markings that could be regarded as art.

Early humans migrated out of Africa and settled in the Middle East. From the Arabian Peninsula they continued to move to South Asia and Oceania, and from there they got to Australia sometime between 44,000 and 50,000 years ago [207]. The evidence from Australia is the first challenge to the idea that artistic behavior began with a big bang of creativity in Europe. Before this purported creative explosion, people in Australia were transporting and processing ochre and decorating themselves and their environment. The earliest findings in Australia show distinct regional styles of engraving and painting. Fragments of possibly painted rocks found in Carpenter's Gap (Tangalma), Kimberley, are from 42,000 years ago, and blocks of hematite with signs of grinding from Malakunanja 2 and Nauwalabila 1, Arnhem Land, are from 40,000 years ago.

Another challenge to the standard art-began-in-Europe story comes from the Blombos Cave in South Africa [208]. This cave is set in a limestone cliff, close to the sea, about 180 miles east of Cape Town. Pieces of red ochre with criss-crossing lines have been found in the cave that dates back to 75,000 and 100,000 years ago and reveal a tradition of abstract geometric engraving. The cave also has polished and ground animal bone tools, dating to 82,000 years ago, making them among the oldest bone tools in Africa. Several stone artifacts from these caves, known as bifacial points, are fashioned in a style that appeared in Europe only 63,000 years later. Besides tools, beads and engravings suggest that the people in these caves liked decoration. They probably gathered shell beads from rivers 20 kilometers away and brought them to these caves. The cave dwellers were selective about their shells, picking large ones to string together and wear. The fact that these people picked shells of similar size and shade, and that the shells have similar patterns of perforations and were worn similarly, suggests that their culture had a developed bead-making tradition.

People in these early days used shell ornaments widely. At the other end of the African continent, perforated shells about 82,000 years old were found in Morocco [209]. These shells also show traces of ochre, of having been suspended, and of prolonged use. People also made decorative shells and beads in the Middle East. Perforated shells have been found at the Near Eastern caves of Qafzeh and Skhul [210]. These shells were transported considerable distances from the sea. They also have ochre remains and show traces of having been strung. The shells from Qafzeh Cave date to 95,000 years ago, and those from Skhul date to between 100,000 and 135,000 years ago. For reasons lost in time, this bead-making tradition that was widespread across Africa and the Middle East died 70,000 years ago.

While the traces found on ancient beads show us that early humans used pigments for decoration, it turns out that pigments were used even before that time. Iron and manganese mineral blocks from 200,000 ago have been found at Twin Rivers, in central Zambia [211]. From these blocks, people made yellow, brown, red, purple, pink, and dark blue pigments. The particular minerals don't occur naturally where they were found, which suggests that people collected and transported them. The people probably did use pigments for practical purposes. For example, ochre is used to make glue to construct tools, and similar pigments were probably used to help preserve wooden tools. But findings from Pinnacle Point in South Africa from 164,000 years ago suggest that humans were interested in pigments for more than their utility. They used highly saturated red pieces of ochre more commonly than other available hues; they may have had a preference for the reddest ochre that went beyond its utility [212].

Is there something unique about the *Homo sapiens* brain that gave these people the capacity to decorate? Probably not. Even earlier than the findings from Africa, Neanderthals were using red ochre as far back as 250,000 years ago [213]. In Tata, Hungary, Neanderthals carved an exquisite oval plaque from part of a mammoth molar. At the same site, a Neanderthal engraved a line on a fossil nummulite that combines with a natural fracture line to make a perfect cross [214]. Neanderthals decorated their graves and adorned themselves with feathers [215]. *Homo heidelbergensis*, a species that predates Neanderthals, may have made geometrically complex and rhythmic engravings on bone and antler artifacts [216].

Among the most striking findings from this period are very old Venus figures. Venus figures are typical of a much later period, around 28,000 to 22,000 years ago, However, two more ancient figures that are not linked directly to the later figures are the Berekhat Ram and the Tan-Tan figure. The Berakhat Ram figure, from the northern Golan Heights in the Middle East, is a basaltic tuff pebble about 35 centimeters long. This piece, probably 230,000 years old, has a natural shape that looks like a head, torso, and arms of a female human. The neck and arms and chest look like the pebble was modified to emphasize its iconic shape [217]. The Tan-Tan quartzite figurine found in Morocco is also a natural object that was fashioned. It is roughly 6 centimeters long. The surface has eight grooves that symmetrically underline the human form of the object. It was coated with red paint made of iron and manganese [218]. As of this writing, its approximate age of 400,000 years makes it the oldest known sculpture.

Cupules are very old rock engravings found all over the world. They are cup-shaped indentations found in very hard rock. If they are decorative, then the oldest known "human art" may be the series of cupules in two quartzite caves in India: the Auditorium Cave in Bhimbetka and the Daraki-Chattan rock shelter [219]. These sites are older than 290,000 years, and may be as old as 700,000. In Auditorium Cave, a large horizontal tunnel roughly 25 meters long leads to a cavernous high-ceilinged chamber with three exits: the whole cave gallery resembles a cross marked in the center by a huge rock 9 cubic meters in size, named "Chief's Rock." This gallery has nine cupules in a large vertical boulder above ground level. A tenth cupule has a single meandering groove close to it. Nobody knows what these cupules mean, why they were made, or why similar indentations are found on every continent except Antarctica.

As we go back still further in time the record gets increasingly murky. As far back as 800,000 years ago, in the Wonderwerk Cave, in South Africa, colored pigments were used. *Homo erectus* in Africa collected quartz crystals, with no obvious practical purpose, 850,000 years ago [220]. Maybe, even back then, people just liked shiny objects. The oldest artifact that might be regarded as an art object is a red jasperite cobble from the Makapansgat Cave in South Africa. *Cobble* is the name geologists give to a rock fragment slightly larger than a pebble. The cobble could not have occurred naturally in this cave, so it must have been carried there. The cobble is worn in a way that makes it look like a crude face [221]. Scholars speculate that it was carried because it had a special significance. If this object did have such significance, then even the ancient *Australopithecus* may have had a rudimentary symbolic capacity 2.5 or 3 million years ago.

From this brief wander backward in time, we find that artistic behavior is not a special activity that sprang uniquely out of the *Homo sapiens* brain. Certainly, *Homo sapiens* showed a level of artistic complexity not found in earlier hominid artwork. Nevertheless, Neanderthals, earlier *Homo erectus,* and maybe even *Australopithecus* showed rudiments of artistic behavior.

They decorated bodies and bones and stones, placed them in burial sites, and carried pebbles and beads as if they had special value. The sheer diversity of materials and imagery discovered is striking, even though only a fraction of these decorations and artworks could possibly have survived. No grand sweeping narrative explains the archeological record [222]. It seems that artistic traditions popped up, were sustained and transmitted for a while, and then died down in different times and places.

As more sites are excavated, we are left without a singular explanation for the full range of Paleolithic art. It doesn't make sense that all these activities were inspired by or met a single function. The anthropologist Margaret Conkey [223] suggests that even labeling the corpus as Paleolithic art reifies the idea that there is a body of work that represents one movement. She says the urge to look for a sequence that goes from simple to complex is a bias of nineteenth-century anthropological thinking. She thinks that art in the Pleistocene period is better framed as diverse and as arising from varied local conditions.

Do scholars agree about anything when it comes to these diverse examples of ancient art? Most scholars agree that the production of these artifacts needed planning, some technological abilities, and a rudimentary social infrastructure. They tend to view these artifacts as being distinct from tools. The artifacts were not made to be utilitarian, at least not in a very direct way. Mostly, the scholars agree that the artifacts represent symbolic behavior. These points of agreement address the conditions for making art, and not what the artworks meant. Why people would have spent hours grinding pigments, collecting the right shells, carving bones and stones, building scaffolds to paint walls is just not known.

Let's return to the amazing cave paintings. What do they mean? Can a focus on these paintings, that everybody agrees are a high point of ancient art, reveal why we make and appreciate art? The answer, unfortunately, is no [223]. In the 1800s, scholars thought hunter-gatherers made this art to while away the time. In the 1900s, scholars thought this art recorded hunting rites and rituals. Some scholars thought the paintings communicated practical information about the animals. Others thought the paintings reflected a preoccupation with the magic needed to bring down these animals, or that the paintings depicted trance-like states induced by shamans. Another group of scholars thought the art represented fertility rites, and others proposed that the animals in these paintings symbolically represented different human clans. In their view, the large-scale compositions were stories of human competition, perhaps ways of marking territory. Other scholars have even postulated that the pattern of animal and sign pairings represent abstract principles, such as male and female dichotomies.

The paintings from southern France and northern Spain are the exception and not the rule when it comes to ancient art. Figurative work from the Pleistocene era is exceedingly rare. Humans produced this figurative art under specific geological, ecological, and demographic conditions. The art in the French and Spanish caves does not represent a step in the progressive development of human culture. Instead, it was an intense and localized phenomenon. Cave art was one of several such episodes in human history. Each episode originated for local reasons and evolved with its own idiosyncratic trajectories.

Why was the extraordinary eruption of cave art restricted to this one part of Europe? The extensive limestone formations in the French Pyrenees and Spanish Cantabrian created underground caves needed to both house and preserve this art. The fact that similar caves in other parts of Europe do not have such art means that the right geology was not enough for humans to produce this art. The right climate was also critical. The southwest of France and the north of Spain, close to the western Atlantic coastline, had summer temperatures about 6°C cooler in the summer and 8°C warmer in the winter than inland regions. This moderate coastal climate produced a characteristic landscape with the flora of the tundra not found near other European limestone caves. The area was open and rich with nourishing low-growing vegetation that attracted herbivores like reindeer, horses, oxen, bison, red deer, and more sporadic species, such as ibex, mammoth, rhinoceros, and wild pig. These herbivores brought predators, like lions, leopards, and bears, in tow. The abundant large herbivores also brought humans to the area. These animals, prey and predator, are the main subjects of cave art. As a result of bountiful food, human populations grew. Rather than continue to move nomadically, they settled in the area. This combination of an environment with abundant resources and a growing population provided the conditions in which the art of these caves could be produced [224].

Why did this amazing artistic tradition, which survived for over 20,000 years, disappear? Toward the end of the last glacial period, about 14,000 years ago, temperatures rose sharply by at least 8–10°C. As a result of this global warming, the landscape changed and became fully forested. The open-country animal species on which humans depended were replaced by new, smaller woodland species. Dwindling resources meant dwindling human dwellings. Human technology and culture became less complex. With the exception of painted pebbles and simple engravings, artwork pretty much disappeared from this part of the world. We simply do not know if other complex artistic traditions that blossomed and then died in the distant past are waiting to be discovered.

We ended our discussion of contemporary art with the hope that looking to the distant past might offer clarification. Clearly, this strategy did not succeed. Art turns out to be a tangle regardless of whether we encounter it in edgy Soho galleries or dank Cantabrian caves. Paleolithic art, rather than giving us a clear picture of what art might be and how we might direct our research efforts, ends up being at least as confusing as contemporary art. There are three lessons to be learned from Paleolithic art. The first is that art at its inception was incredibly diverse. This art resists being categorized into one class of objects. Second, when we try to interpret ancient art, we hit a wall. We may admire the art, but that does not mean we understand it. Third, even back then, local demographic and ecological conditions shaped the production and, presumably, the appreciation of art.

This trip back into the worlds of Paleolithic art sets us up for the question driving the next few chapters. If art-like behavior existed as far back as we can see, do we have an art instinct?

Chapter 7

Evolving Minds

How did our minds end up producing these curious objects that we call art? In previous sections, we saw that evolution used pleasure to shape our sense of beauty. Before addressing the question of how evolution shaped our minds to make art, it is worth taking another look at how our minds evolved.

Darwin appreciated that his theories of evolution in biology were also relevant to psychology. In 1966, George Williams wrote *Adaptation and Natural Selection*, a book that many scholars consider to be the first to highlight adaptations as the crucial unit of analysis for evolution. The major criterion for deciding that something is an adaptation is "evidence of design." The term *design* does not imply a sentient designer, as assumed by creationists. It means that adaptations were designed to solve past environmental problems. In 1992, an influential book, *The Adapted Mind: Evolutionary Psychology and the Generation of Culture*, edited by Jerome Barkow, James John Toobey, and Leda Cosmides, provided a collection of papers heralding this nascent field. Together these scholars postulated that investigating our minds in the context of our collective evolutionary history would give us new and fundamental insight into human nature. As I mentioned earlier in this book, evolutionary psychology promises to tell us not just how we are but why we are the way we are.

Will evolutionary psychology answer the question about whether or not we have an art instinct? What exactly do we mean by the word *instinct*? The word is used widely, but often without precision. In animal studies we can distinguish between automatic and learned behaviors. If an animal does something without learning how and is not aware of its actions, and the action is shared by most others of its species, we call the behavior an instinct. Instinctual behaviors can be complex, as seen in the dance of bees that signals the direction and distance of food sources to other bees. Courtship behaviors of animals are also instincts. The meaning of the word *instinct* for human psychology is less clear. In general, behaviors that are stereotyped, seem preprogrammed, and do not need to be learned probably qualify as instincts. So behaviors that we are inclined to do automatically and seem universal in humans are candidate instincts. Most people think of instincts as being hard-wired. For our discussion, I will regard instincts as synonymous with psychological adaptations. Psychological adaptations are complex behavioral patterns built into our minds over many generations. They are designed to enhance reproduction by solving past environmental problems, and they are shared by most humans.

Darwin described evolution as descent with modification. He meant that complex and quite well-designed systems evolve over time. These systems could be organs, such as eyes and livers and brains, or organisms, such as birds and bees and humans. Evolution uses three main ingredients in its recipe: variability, heritability, and selection. *Variability* refers to differences in the properties or behavior of people in a population. The relevant variability for evolution is the kind that enables people with the property in question to survive and reproduce more often than those who do not have this property. A property that varies among people, such as hair color, is not relevant for survival, but a property like the ability to fight off infections is. These properties and behaviors have to be *heritable*, meaning that people biologically pass them on to next generations. Darwin did not know that genes were the vehicle for heritability, but he recognized that some such mechanism must exist. *Selection* refers to the fact that some genes pass on to subsequent generations more easily than others. You may recall my fanciful hoodoo analogy for conveying how passive erosion of physical features could eventually make animate hoodoos agree on their judgments of physical beauty. Selection can also be thought of as a sieve, through which some genetic combinations of an original mixture pass more easily than others. Over many generations, just a slight increase in what passes through makes a big difference in the final proportions of genes and traits and behaviors that we find in the population.

We should be clear that most genetic mutations hurt rather than help us. But occasionally a mutation produces a selective advantage. Over many, many generations, rare advantageous mutations survive and become more common in a population. Through accretion, our physical organs, like the pancreas and the liver, got cobbled together into incredibly complex coordinated systems designed to digest food and filter toxins. Better and safer nutrition then enhanced the organisms' survival, and better pancreases and livers got passed onto later generations.

The fundamental insight of evolutionary psychology is that nature, through evolution, sculpts our brain/mind, just as it does our physical body. The same forces that select genes for physical features of the body also select genes for features of the brain to carry out functions that ultimately give the organism a reproductive advantage. Mental functions that gave our ancestors an advantage in surviving and having children were passed on and accumulated in humans over time. These functions, such as knowing how to find nutrition, select healthy mates, and navigate through difficult terrains, got built into the structure of our brains, piece by piece. Other complex cognitive abilities, such as categorizing and reasoning, counting, recognizing emotions, inferring beliefs and desires of others, and acquiring language to communicate, also gave our ancestors a selective advantage and were passed on to us. Brain adaptations that helped "solve" environmental problems of the past gave our brains its current structure. Evolution has no master plan for our brains. Yet, the persistent tinkering means that mental functions work pretty well, even if not perfectly.

Evolutionary psychologists face a "reverse engineering" problem. To understand the mind, they work backward from what we observe today to figure out the important psychological functions that were designed by pressures of the past. They also have to consider other evolutionary byproducts that came along for the ride, and plug-ins that were added to our minds by recent local conditions. Dissecting our mental anatomy to figure out which parts are adaptations and which are not is not always easy. We are out of touch with the environments in which our brains evolved. Those environments changed over very long periods, predating and continuing through the almost 2 million years of the Pleistocene. Over this huge expanse of time, different environmental pressures worked their selection magic on roving bands of tens to a few hundred early humans. For example, being able to drive safely in multilane highways or to figure out complicated health care insurance coverage schemes (in the United States) certainly enhances our chances of survival. But our ancestors did not need these skills as they evolved. We can only guess at the environmental pressures that selected traits with reproductive advantages that make up our modern minds.

Our brains and bodies are more than a collection of adaptations. Our brains are also a grab bag of other mechanisms that came along for the ride, got modified, and even did their own modifying. Examples from biology illustrate the way these mental mechanisms and physical properties get collected. Bones are made of calcium salts. Calcium was selected presumably because its structural properties were better than those of other materials available to early organisms. Calcium salts are also white. The fact that bones are white is a byproduct of the selection of calcium salts. Their whiteness has no functional significance. The evolutionary biologist Steven Jay Gould referred to these kinds of byproducts as spandrels. As I mentioned earlier, architectural spandrels are the spaces between arches produced as a by-product of columns and arches. Spandrels are not structurally significant themselves, but they can be used decoratively.

Evolutionary by-products can become useful as the environment changes. Then they are called exaptations [225]. Most evolutionary biologists agree that feathers first served to trap and preserve heat in birds. This property of feathers is the reason down jackets and blankets keep us so warm. When early bird populations faced pressures to fly, feathers that were adapted to trap warmth became useful for another purpose. They got exapted to fly. The tail and wing feathers got further modified to be stiffer and asymmetrical around the vane, resulting in a better aerodynamic design that made flying more efficient. These feathers were now secondarily adapted. So exaptations are features that did not start out useful. They become useful when the environmental pressures changed. Secondary adaptations are selected modifications of exaptations that make them even more useful.

Adaptations, spandrels, exaptations, and secondary adaptations are all biological and psychological mechanisms that collect within us over many generations. Long-term environmental pressures fashion them, but that is not the whole story of the evolution of our brains. Local conditions also tinker with us.

Consider a document stored in your computer. The document has some kind of binary code that most of us don't understand. Every time we open the document, barring corruption of our computer's hardware or software, the stored code expresses itself the same way, and the same document shows up on our screens. Wouldn't it be bizarre if the same binary code produced different documents on our screens when the room in which we were working changed? Sometimes, genetic codes behave bizarrely like that. Genes can express themselves differently because of local conditions. Take the example of caterpillars of the moth *Nemoria arizonaria.* These caterpillars develop on oak trees. The caterpillars that hatch in the spring eat oak flowers and then develop to look like the flowers. The caterpillars that hatch in the summer eat the oak leaves and then develop to look like twigs. This difference in diet makes the same genes produce very different physical bodies [226]. The ability to produce different bodies is adaptive because it provides better camouflage for different seasons. Another example comes from under water. Sea slugs eat a sea moss called bryozoans. Bryozoans detect a chemical exuded by sea slugs. When they detect the chemical, bryozoans grow protective spines. Otherwise, they do not. The same genes produce dramatically different body shapes depending on the signals they sense in their environment [227].

Local environments can also make genes produce dramatically different behavior. The African cichlid fish, *Haplochromis burtoni,* has two kinds of males. One is territorial and brightly colored. This male has developed testes, reproduces, and protects its territory aggressively. The other kind of male is not territorial. This male looks bland, has undeveloped testes, and swims with females. Predators are more likely to eat the territorial male, attracted by his colorful swagger. If a territorial male dies and a non-territorial male takes possession of the abandoned area, within days the non-territorial male becomes brightly colored and develops mature testes and behaves aggressively. If a territorial male is displaced and he can't capture a new territory, he loses his bright colors and his testes atrophy [228]. These environmental changes happen rapidly and unpredictably and trigger a cascade of events in genetic expression that produce striking changes in the organism [229]. Environmental pressures can have a dramatic impact on how organisms look and behave over a much shorter time period than what evolution needs to sculpt its adaptations.

Organisms also change their environments. For example, beavers build dams, moles dig burrows, and humans grow food. They change their environments to create local niches. Each of these niches changes the selective pressures faced by later generations. With the development of human agriculture, more people had more food using less land. Population densities grew. The switch from a mostly protein to a stable starchy diet resulted in malnutrition, and the higher population density along with the domestication of animals allowed infectious diseases to spread quickly [230]. Thus, humans created new environmental niches that changed selection pressures that later humans encountered. Environmental niches can be created very locally, as the following example shows. Some West African populations cut clearings in the rainforest to cultivate yams. As a result of more standing water, more malaria-carrying mosquitoes bred. Malaria, in turn, increased the frequency of the gene for sickle cell anemia in the population, because this gene offers protection against malaria. So cutting down trees for more stable sources of nutrition resulted in more people having sickle cell anemia [231]. When we modify our environment, these modifications can bounce back to influence our biology and psychology.

In this chapter, we have seen how evolutionary mechanisms play out in biology. These examples give vivid form to the analogous dynamics that must also play out in our psychology. Like our bodies and their physical features, our minds evolved to contain different mental mechanisms, some useful and some just left over because getting rid of them was not worth the effort. Adaptations in the collection helped our ancestors survive and reproduce. Most adaptations are still useful now, but some are not because they are no longer relevant to survival and reproduction in the modern world. Spandrels came along for the ride and serve no specific function in and of themselves. Spandrels become exaptations if changes in the environment make them useful. Exaptations themselves can be modified by selective pressures to become secondary adaptations. For the purposes of our investigation of aesthetics, we are most interested in how these mechanisms accumulated in our minds over many years. We gather other mental mechanisms in response to local environments. Sometimes we create local environmental niches that in turn influence our bodies and brains. With this collection of mechanisms in mind, we are ready to examine the evolution of art. Which of these mental mechanisms lets us make and appreciate art?

Chapter 8

Evolving Art

Do we have an instinct for art? Art lovers have the powerful intuition that art is a deep and integral part of us. The impulse to make and enjoy art feels fundamental to our nature. It is a short step from that powerful intuition to think that if something is fundamental to our nature, surely it must be an instinct inserted by evolution. This thought is further bolstered by the observation that we see art wherever we look. Art, it would seem, is universal. If a behavior is universal, surely it serves an adaptive function. Again, it is a short step to believing that we have an art instinct.

At the other extreme from the position that we have an art instinct is the view that art is simply a by-product of other adaptations. Several intellectual heavyweights endorse this view. Steven Jay Gould and Richard Lewontin [232] argued that human culture as we know it has only been around for about 10,000 years, not long enough for the brain to have changed in a substantial way from selective pressures. Given this limited time frame, cultural artifacts must be a by-product of our large brains that evolved to solve other problems faced by our more ancient ancestors. The psychologist Steven Pinker famously likened art to cheesecake. Using the example of music, he suggested that music is cheesecake for our ears. Cheesecake is an artificial by-product designed purely for our pleasure and plays on our needs for fat and sugar. Analogously, music plays on other adaptive needs, like emotional processing, auditory analyses, language if it has lyrics, motor control in so far as music is associated with dancing, and so forth. For Pinker, music and other arts came along as a delightful by-product of mental faculties that evolved for other adaptive purposes.

Comparing the two positions, we have the art-as-by-product view, that art serves no real purpose outside the pleasure it gives us, and the art-as-instinct position, that art does have purposes and these purposes are adaptive. Let's examine the arguments for art being an instinct in more detail through the writings of the independent scholar Ellen Dissanayake, the evolutionary psychologist Geoffrey Miller, and the philosopher Dennis Dutton. After establishing the arguments for an art instinct, we will test the adequacy of these arguments as applied to art.

Ellen Dissanayake did much of her thinking and writing outside of the hallowed grounds of the academy. She sat next to me during dinner at a 2009 neuroaesthetics conference in Copenhagen. A quietly intense woman, she told me that her key to academic success was to write about something nobody else cared about at the time and then to persist in developing the initial ideas. Being an outsider allowed her to raise important questions in aesthetics that had not been asked. She was prescient in linking art to evolution. She developed her evolutionary aesthetics in the early 1980s, well before such thinking was popular among academics. Appealingly, she approaches aesthetics with a broad cross-cultural view and is not tethered to theorizing about Western art, unlike much of aesthetic scholarship.

For Dissanayake, art is embedded in rituals. She shifts the emphasis in analyzing aesthetic experiences from an individual's encounter with art to the social roles played by these encounters. People engage with art in ways that foster cooperation. She calls this engagement "artification" or, more colloquially, "making special." Using ordinary objects ritually makes them special. We simplify, stylize, exaggerate, and elaborate objects to make them special. The intensity with which we then repeatedly engage them distinguishes these special objects from everyday objects.

Dissanayake offers several reasons for art being an instinct [233]. She starts with the general proposition that art is universal. After all, children engage in the arts spontaneously. They scribble, sing, move to music, fantasize, and play with words. She also thinks that art gives pleasure. For Dissanayake, appreciating art is like spending time with friends, having sex, and eating good food. It serves a basic human need.

After making these general claims about art as an instinct, Dissanayake hones in on two specific reasons for "artification" being adaptive. First, a society is more cohesive if it has communal rites that bind its members together in common beliefs and values. Artification is ritualized behavior that binds people together, and a cohesive group is better equipped to survive ancestral pressures than loosely connected groups of individuals. Her second argument links artification to the evolution of mother–infant bonds [234]. Becoming bipedal meant that *Homo erectus*' pelvis had to narrow. However, while their pelvises narrowed, early humans were also evolving larger brains and skulls. These changes created the problem of mothers with narrow pelvises having to bear big-headed babies. Specific physical adaptations dealt with the problem. Females evolved pubic bones that separate during childbirth, and infants evolved skulls that are compressible. Pregnancy periods also got shorter, which meant that the infants' brains and bodies had to continue to mature after birth. Thus, human infants are more dependent on their caregivers for longer periods of time than is typical of other primates. Along with these physical adaptations, Dissanayake posits that specific rituals became critical for the survival of the infant. Mothers developed behaviors, such as smiles, head bobs, eyebrow flashes, soft undulating vocalizations, repetitive pats, touches, and kisses, that strengthened the social bond with their infants. These ritualized behaviors fall squarely into the behavioral patterns of simplification, repetition, elaboration, and exaggeration that Dissanayake thinks make up artification. Mothers make infants special. These ritualized behaviors are at the root of making other objects special. They are at the root of making art.

Dissanayake's version of art as an instinct is a kinder and gentler account of evolution than Geoffrey Miller's proposal. Miller emphasizes the costly signal hypothesis that is used to explain evolutionary psychologists' favorite example of nature's extravagance, the peacock's tail [235]. The drama of sexual selection encourages males, competing for choosey females, to advertise their superior fitness by indulging in frivolous displays like the peacock's tail or the deer's antlers. As we saw earlier, these displays are costly because they paradoxically work against natural selection. The tail hampers the peacock's movement, making it more vulnerable to predators. But here the advantages of sexual selection trump the disadvantages of natural selection. These showy birds attract more mates and have more offspring despite being at greater risk of dying young. For Miller, art is the human counterpart of costly signals found in animals. It is rooted in strutting males producing useless adornments as they compete with each other trying to convince female viewers that their paintings are bigger and better than the next guy's paintings.

The late philosopher Dennis Dutton makes his opinion clear from the title of his book, *The Art Instinct.* His position is a combination of Dissanayake's social-cohesion argument and Miller's costly display argument. Dutton regarded art as having a cluster of features, none of which are necessary or sufficient to define art. Despite this inherent variability, he suggested that art must derive from a natural, innate source, because art practices are easily recognized across cultures. Based on a discussion of the common attributes of landscapes that we find beautiful, he thought we share an instinct for beauty that is relevant to art. He dismissed the possibility that art is a by-product of evolution because he thought by-products couldn't possibly be relevant to our lives.

Dutton thought evolution generated an art instinct in two ways. First, we evolved creative capacities to survive the hostile conditions of the Pleistocene era. By inventing and absorbing stories, early humans worked out "what if" scenarios without risking their lives. They could pass on survival tips and build the capacity to understand other people and situations. The best storytellers and best listeners had slightly better odds of survival, giving future generations more good storytellers and listeners. Presumably, these creative capacities generalized to other forms of art. Second, the creative types would have had better luck at wooing mates and reproducing. Here, he endorsed Miller's costly signal sexual-selection argument. For Dutton, art is an instinct fashioned by both natural and sexual selection.

Each of these scholars thinks we have an instinct for art, but their views on exactly how this instinct evolved varies. Embedded within their views are four ideas. Let's examine each of them: (1) art is the expression of an instinct for beauty, (2) art is a costly signal advertising reproductive fitness, (3) art is useful, and (4) art promotes social cohesion. To examine these ideas, I will modify an approach from medicine used to assess diagnostic tests. A test for any condition or disease, such as Alzheimer's disease, has its own sensitivity and specificity. *Sensitivity* refers to how often the test is positive if the person actually has the disease. Insensitive tests often miss the disease. *Specificity* refers to how often the test is negative when the disease is not present. Non-specific tests are positive for other diseases or when there is no disease at all. The best tests are highly sensitive and highly specific. Analogously, let's assess whether we have an art instinct by evaluating the sensitivity and specificity of these claims. The more sensitive and specific the claim, the more convincing it would be.

Does art represent an instinct for beauty? In the first section of this book, we discussed ways in which our minds adapted to beauty. We evolved to find certain faces, bodies, and landscapes beautiful, because the trait of experiencing pleasure in these "beautiful" configurations provided a reproductive advantage through sexual or natural selection. But to say a pretty face or a beautiful landscape is the point of a painted portrait or landscape is to reduce art to visual candy. While most people link beauty to art, beauty is neither sensitive nor specific to art. The lack of sensitivity is obvious enough. Recent conceptual art can be shocking or provocative without being beautiful. This divorce of art from beauty is not simply a recent phenomenon. Earlier artworks from masters like Francis Bacon, Edvard Munch, Francisco Goya, and Hieronymus Bosch were powerful without necessarily being beautiful. The lack of specificity of the argument that art represents an instinct for beauty is obvious. There are plenty of beautiful objects, like people and places and flowers and faces, that are not art. Having instincts for beauty does not mean that we have an instinct for art.

Is art an instinctual display of a costly signal? The idea of art as an individual's extravagant display comes from an eighteenth-century conception of art. In his book, *The Invention of Art,* Larry Shiner describes how eighteenth-century European theorists began to distinguish art from craft. This distinction did not exist in the public imagination before then. A growing middle class and newly developing art markets adopted these theorists' view and started to think of art as an expression of an individual's creative genius. This conception disregards older examples of art that were typically produced in the service of a patron or church or state. Those commissioned artworks were fundamentally functional in nature and not swaddled by an admiration for the artist's skill. For example, medieval Christian faithfuls gazed at images like the twelfth-century Russian icon, *Our Lady of the Tenderness,* as part of their spiritual practice, without thinking about the person who made the inspiring image. The costly display argument is not sensitive to the rich history of these art traditions. The costly display argument is also not very specific. As I mentioned before, watches that cost tens of thousands of dollars, cars that cost hundreds of thousands of dollars, and homes that cost many millions of dollars are costly displays. Undoubtedly, some men use these signals very effectively to attract women. But do we want to call conspicuous displays of consumption art? A costly display argument is not particularly sensitive or specific for art.

Is art so useful in our lives that it must be an instinct? This argument also doesn't work. Being useful now is not a measure of whether something evolved as an instinct. Adaptations evolved to be useful in our ancestral past. Ironically, sometimes we are more confident that behaviors are adaptations when they persist despite not being useful. For example, the pleasure we get from sugar and fat is an adaptation from a time when we couldn't easily satisfy these nutritional needs. This same instinctual pleasure is a disaster in economically developed countries, where easily available cheap sweet and fatty foods create epidemics of obesity and diabetes. Paradoxically, the adaptive value of this trait from an earlier era still annoyingly guides our behavior. So being useful in our lives now is not a sensitive way to establish that something, be it the love for art or ice cream, is an instinct. How specific is the usefulness argument? Here, we can return to the example of written language. As I argued earlier, written language is a prime example of something that is not an adaptation and is integral to most of our lives. So being useful in our lives now is not very specific in making the case for an instinct.

Finally, is art an expression of our instinct for social cohesion? While social cohesion and the strengthening of bonds can be an important function of art, it is quite clear that not all art serves this purpose. Philosophers and cultural theorists tell us that art can be defined in different ways that serve different functions. Imagine a lone artist toiling away in isolation, pushing his or her own boundaries, producing works not seen by others. Would these works not be art because they were not communally appraised? Conversely, actions that create strong social bonds and make objects special are not specific to art. Sports teams and their fans bond with repetitive, exaggerated, and elaborate behaviors. Most of us do not regard football jerseys as art and hooligan behavior as artification. Beyond sports, militaries around the world use ritualized behaviors to create cohesion. Would we say that goose-stepping soldiers waving martial banners are engaged in artistic behavior?

Despite these reasons for doubting the idea that we have an art instinct, many of you might not be convinced. The idea still feels like it contains a germ of truth. You might think it premature to abandon the search for a unifying instinctual theory of art. You might even have noticed that I slipped between talking about *explanations* for the existence of art to talking about *definitions* of art. My strategy of scrutinizing the sensitivity and specificity of the idea that art is an adaptation is similar to trying to identify necessary and sufficient conditions that define art. Twentieth-century philosophers realized that that strategy simply doesn't work to capture all art. When I show that adaptation arguments are not sensitive or specific to art, maybe I am saying that adaptations don't work to define all art. You might reasonably object that adaptations are about why things came to be, not about what they are. Maybe adaptations succeed better as explanations for art, or at least as explanations for the origins of art. Perhaps we should entertain a more modest proposal than the instinctual theory of art. Perhaps adaptations contributed to the original creation of art. The question of whether specific adaptations sustain art would have to be examined on a case-by-case basis. Maybe we do not have to completely abandon the instinctual theory of art. Maybe we do not have to resign ourselves to the art-is-a-by-product view. Maybe there is a third way to think about art. In the next chapter we will consider this third way through the example of a little bird bred in Japan.

Chapter 9

Art: A Tail or a Song?

If art is not an instinct, how do we explain the fact that we are surrounded by art? How do we explain the fact that rudiments of art exist as far back into the past as we can see? The belief that art must be an expression of a deep instinct in our collective psyche is hard to shake. At the same time, the sheer variety of art cannot be ignored. We cannot be blind to the fact that art is shaped profoundly by history and culture. Art can be an object of contemplation or of reverence just as easily as it can be a commodity buoyed by institutional and market forces. When we emphasize the universality of art, we slide into thinking of art as an instinct. When we acknowledge the sheer diversity and cultural fashioning of art, we slide into thinking of art as a spandrel. Is there a third way to think about art?

Is art more like a peacock's tail or like a Bengalese finch's song? We haven't talked about the Bengalese finch's song yet, but the question is another way of asking if art is the expression of a finely honed adaptation like the peacock's tail, or if art is an agile response to local conditions like the finch's song. We saw that some caterpillars develop different bodies depending on what they eat; that some sea moss change dramatically when they sense chemical signals in their environment; that some fish change their appearance and behavior if they accidentally inherit or lose a territory to protect. I used these examples to show that organisms can change quickly and dramatically in response to local environmental conditions. These changes occur over much shorter periods of time than evolutionary adaptations that accumulate over long swaths of time. To see how evolutionary responses to long or short intervals of time might relate to art, let's turn to the peacock and the finch.

The peacock's tail, of course, is the evolutionary psychologist's favorite example of a costly display that advertises the bird's fitness. The tail is elaborate and beautiful. The tail also makes it harder for the peacock to move quickly, leaving it vulnerable to predators. Sexual selection, rather than natural selection, drives the development of these colorful tails. Many cultural artifacts are thought to be like the peacock's tail. For some scholars, as we saw in the last chapter, art is a prime example of a costly display. Elaborate, beautiful, and not very useful certainly sounds like a lot of art. As I argued in the last chapter, this view of the evolution of art is not very satisfying.

We need a different example from biology to capture the evolution of a behavior that is complex, elaborate, and varied. In addition to being varied, the behavior is unpredictable and its content responds to its local environment. After all, a hunk of lard is art only under the right cultural conditions. The song of the Bengalese finch gives us a useful example. The song, rather than being driven by ramped-up selective pressures as with the peacock's tail, emerges from a relaxation of these same pressures. In general, the relaxation of selective pressures puts a limit on adaptation and promotes variability in biological organisms [236]. The biological anthropologist Terence Deacon suggests that the social use of language and many of our cultural practices emerge when selection pressures relax [237].

The Bengalese finch is a domestic bird bred in Japan. It descended from the feral white rumped munia, which lives in the wild throughout much of Asia. Male munias, like many birds, sing a stereotypic song to attract mates. Japanese bird breeders mated the munia for its plumage to produce birds with especially colorful feathers. In this artificial niche, over 250 years and 500 generations, the wild munia evolved into the domestic Bengalese finch. The domesticated birds' singing abilities are now irrelevant to their reproductive success. While they were being selected to be colorful, their songs, rather than withering to a croak, became more complex and more variable, and the sequence of notes became more unpredictable [238]. The Bengalese finches also became more responsive to their social environment. They can learn new songs more easily than their munia ancestors and even learn abstract patterns embedded in songs [239]. A Bengalese chick can learn a munia's song, but a munia chick can only learn the song it is destined to sing. According to Deacon, as the content of the song became irrelevant to usual selective pressures (identifying the same species, defending territories, avoiding predators, and attracting mates), the natural drift and degradation of genes that program the stereotypic song could occur. The contaminated genes allow for neural configurations that produce songs that are less constrained and easily perturbed. What the Bengalese finch hears in its environment increasingly influences the content of its song.

The changes in the finch's song are accompanied by interesting changes in its brain. The neural pathways for innate songs in the munia are relatively simple and mostly controlled by one subcortical structure called the nucleus RA. By contrast, the neural pathways for the Bengalese finch's songs are widely distributed across the cortex and come online more flexibly. Different parts of the bird's brain now coordinate the output of the nucleus RA [240]. The difference between the munia and the finch's song, by analogy, is that of music played in a prescribed manner versus music that is improvised. As genetic control over brain function got looser, instinctual constraints on the bird's song got less specific. The finch's brain became more flexible and its behavior more improvisational and responsive to local environmental conditions.

Thus, *opposite* evolutionary forces drove the emergence of the peacock's tail and the Bengalese finch's song. Ramping up selective pressures produced the tail, while relaxing these same pressures produced the song. The song started as an adaptation but evolved into its current form in a relatively short time, precisely because it no longer served an adaptive function. The art we encounter today is more like the Bengalese finch's song than the peacock's tail.

Art is like the finch's song both in its biology and in its intrinsic characteristics. Our art experiences are coordinated in the brain by widely distributed neural ensembles. We do not have a unique art module in the brain. When we engage with art, we use systems dedicated to sensations, emotions, and meaning in other contexts. The specific systems engaged during any one encounter vary depending on the kind of art we are perceiving or creating. This flexible setup of brain structures to coordinate complex behavior is similar to what we see in the finch's brain when it sings. There is no specific song module in the finch's brain. Rather, different cortical structures flexibly fire to coordinate the song the finch happens to be singing.

Art is complex, that is a given. Art is also highly variable, which is why different pieces of art can look nothing like each other; we even have trouble clearly defining which objects are art. Art is also exquisitely responsive to its local cultural environment. A Paleolithic painter might be preoccupied by his quarry, a medieval Christian acolyte by the Holy Mother, a Renaissance artist by his patron, and a contemporary artist by her social cause. Analogously, the finch's songs are complex, as are many bird songs. Unlike many bird songs, the finch's song is variable. The same bird may sing variations on its song in different contexts, and different finches learn to sing different songs. The birds' exquisite sensitivity to their environment is reflected in the content of their songs.

The Bengalese finch's song, in all its variable glory, is still rooted in the white rumped munia's instinctual song. Art, like the finch's song, has adaptive roots. The capacity for imagination, the ability to use symbols, the feeling of pleasure from beauty, the predisposition to social cohesion, may very well be at the root of art production and appreciation. However, these roots are removed from many present-day encounters with art.

The capacity for imagination and the ability to symbolize are preconditions for art. We use these capacities to make and appreciate art. As we saw in the last chapter, art could be the expression of an adaptation by serving as a vehicle for beauty or social cohesion. However, art today can express beauty, but it need not. Art can promote social cohesion, but it need not. Untethered from the adaptive advantages of beauty or of social cohesion, art can become more variable. Art makes use of its adaptive roots, but its current power comes from its flexible and ever-changing nature. Contemporary art is formed in local environmental niches made by humans, and rather than being controlled tightly by an instinct, it blossoms precisely because it is untethered from these instincts.

Let's consider one possible objection to the idea that art blossoms precisely when selective pressures are relaxed. What about revolutionary or dissident art? Such art, often extremely powerful, is produced under great duress. To take a recent example, murals and graffiti art were vivid forms of expression in Tunisia and Egypt during the 2011 Arab Spring protests. The Egyptian street artist Ganzeer declared, "Art is the only weapon we have left to deal with the military dictatorship" [241]. It would seem that fear for one's life can inspire art, maybe even extraordinary art.

Revolutionary art does not undermine the idea that art blossoms when selective pressures are relaxed. Before developing this argument, it is worth emphasizing a critical characteristic of increased or relaxed selective pressures: variety and what happens to that variety. Variety in behavior (phenotypic variability) is a product of variability in genes and the environment. Random mutations that produce genetic variability typically do not help the organism. That is why so many genetic mutations are fatal or produce disease. Typically, genetic mutations get weeded out. The evolutionary engine that does the weeding out drives behaviors toward uniformity, toward what evolutionary biologists call "fixation." When adaptive behaviors are relaxed from selective pressures, deviant behaviors do not need to be eliminated. The behavior no longer matters for survival. It is free to "mutate." So, in the case of increased selective pressures, variety is culled; in the case of relaxed selection, variety blossoms.

Consider the complex set of behaviors that promote social cohesion. As we saw in the last chapter, art production and perception is one (but by no means the only) such behavior. Behaviors that promoted social cohesion among groups of a few hundred Pleistocene wanderers probably do not have the same force in our complex society as they did back then. We have offloaded some of the advantages of socially cohesive individual behavior into laws and rules enforced by "authorities." As the constraints on individual social behaviors diminish, acts that arose to be socially cohesive can drift. Art as an expression of social cohesion can change as pressure for art to do the work of cohesion matters less. This new openness and variety in art can persist as long as countervailing forces do not weed it out. The same dynamic plays out in each of the adaptive functions proposed for art. If the selection pressures that gave rise to these adaptations relax, the behavior is free to drift. The analogy between the finch's song and art makes sense because the structural dynamics underlying both behaviors are similar. Both the bird's song and art start with adaptive functional purposes. They are then either honed by or released from environmental selective pressures. When honed, they become stylized and exaggerated. When released, they become more varied.

The alternate dynamics of selection and relaxation means that art is sometimes sculpted by selective environmental pressures (imposed by a cultural niche), and it is sometimes free to change if those pressures relax. When the former, art is stylized and changes are gradual refinements occurring within a narrow range of form and content. Medieval Christian iconography might be an example of this kind of art. The functional role that such art played in promoting social cohesion within the church meant that changes to art were incremental and refined to enhance those functions in this particular environmental niche. The art one might see on a tour of different medieval Christian churches would be quite restricted in style and content, a far cry from the range of what one might see on a tour of museums of modern art.

To return to revolutionary art, in oppressive regimes, artists who step out of line can be imprisoned or put to death. The selection pressure on art production is severe. But regimes change. Revolutionary art emerges when oppressive regimes show signs of losing their grip on their people. The art of the Arab Spring exploded precisely when change was in the air and selective pressures were starting to relax. Revolutionary art accelerated changes that were stirring.

Based on the dynamics of increased or relaxed selection pressures, I predict the following. Severely oppressive conditions that persist over long periods of time would prevent the emergence of art that is varied and looks creative to our modern eyes. Art in these societies, if produced at all, would be stereotyped, ornamental, operate within narrowly prescribed rules, and probably serve as state propaganda. My guess is that not much art that we would regard as creative is now being produced in North Korea. If and when North Korea opens up to much of the world, we will not find a treasure trove of art produced by creative artists struggling in isolation against impossible odds. However, when the selection pressures of an oppressive regime relax, during periods of revolution, creative and varied art will seep out. The Internet now provides such an outlet, a release from societally imposed selective pressures. For example, the opening of China emboldened revolutionary artists in this transition phase. The Chinese dissident artist Ai Weiwei, who was one of the designers of the Beijing bird nest Olympic stadium and recently named the most powerful art figure by the magazine *Art Review*, has made use of the Web as a vehicle for his art that protests oppressive state policies [242]. Even when placed under house arrest, he set up Webcams in his home, so that the visual record of his confinement itself became a form of dissident art.

My second prediction is that after a revolution has succeeded, resulting in the state changing from being oppressive to open, the nature of the art also changes. Revolutionary art occurs in transition between selective and relaxed pressures imposed by the environmental niche, which in this example is the state. Once the state allows individual freedom, we see the wide variety of simultaneous art practices that emerge in New York, Paris, Barcelona, and any major city in an open society. If the logic of this argument is sound, then the variety of art in a society at any particular time is a measure of its level of freedom. The more the state applies selective pressures on its artists, the more stylized and limited the range of art produced in that culture. The pressures need not be practices of an oppressive regime of the kinds we have been discussing. The pressures may simply be tough economic times in which art behavior is selected and confined by financial forces. The more the arts are released from selective pressures, whether they are state oppression or economic deprivation, the more the arts in that culture are free to vary.

My predictions relate back to the finch analogy. The point of the analogy is not that the finch's song is art and the munia's song is not art. Rather, both songs serve as examples of the qualities of art that emerge in a given environmental niche. The munia song, like state-controlled art, is not as variable as the finch song. Conversely, the relaxation of selection pressures on bird songs and art increases the variety of options available to the community. While art can be an expression of an instinct, it often is not. In fact, art that we might regard as most unpredictable and innovative arises precisely under conditions of relaxed selective pressures.

We started this chapter in search of a third way to think about art. We need this third way to thread between the two traditional ways of thinking of art as either an evolutionary by-product or as an instinct. Otherwise, we can't account for the sheer variety of art and at the same time account for its universality. The white rumped munia and the Bengalese finch give us this third way, by showing that a behavior can have adaptive roots and then evolve when selective pressures on the adaption are relaxed. As we saw in the discussion of revolutionary art, art changes depending on its environmental niche. It can be finely honed to serve a purpose. It can be released from the burden of serving a purpose, mutate unpredictably, and blossom even to exist simply for its own sake. Art can be both the expression of an instinct and a relaxation from this instinct. The key is whether art in a specific cultural environment follows narrowly prescribed rules or whether it is varied and unpredictable. Art, it turns out, signals our freedom.

Chapter 10

The Serendipity of Art

I introduced this book by describing my walk to the Museum of Modern and Contemporary Art in Palma, Mallorca. I took in the beauty of the bay, enjoyed the mastery of Picasso and Miró, and was confused by the contemporary art exhibit called "Love and Death." Let's return to this scene after having wandered through the science of beauty, pleasure, and art.

The Bay of Palma was beautiful. The glistening water and the waving palm trees gave me pleasure. At the time, I thought that most people would find the scene beautiful. This intuition was probably correct. We have an instinct for beauty. More accurately, we have instincts for beauty. Several adaptations make us find some objects beautiful. Beauty turns out to be a mongrel. When we considered facial beauty, we saw that averaged faces signal greater underlying genetic diversity, exaggerated sexual dimorphism advertises fitness, and symmetry both represents fitness and allows easier processing of any visual object. When we examined landscapes, we found that a different collection of adaptations make scenes more or less attractive. These adaptations signal sources of nutrition and protection from danger. Finding beauty in abstract objects, such as mathematical theorems, relies on yet other adaptations, such as the ability to reduce complex information, into understandable nuggets. Our general experience of beauty is the result of a loose ensemble of evolutionary adaptations. What ties these people, places, and proofs together into the experience of "beauty" is that our ancestors, who happened to find pleasure in these objects, were the people who had more children. Most people would probably find the view of the Bay of Palma beautiful because most people inherited the same ancestral pleasures.

Reproducing a beautiful scene, like the bay of Palma, does not automatically make great art. Most photographs or paintings of beautiful scenes, like beach sunsets, look trite. We noted that aesthetics and art are different. We have aesthetic responses to any number of objects that need not be art objects. Art objects arouse reactions that can be pleasurable, but they need not be. Even though most people associate beauty with art, art need not be beautiful. Often, it is not.

Looking at the Picasso plates and the Miró prints in the museum gave me great pleasure. Our pleasures, regardless of their sources, funnel through the same brain systems. As we saw in the last section, our core pleasures work through the ventral striatum that is embedded deep in our brain. The fact that different pleasures funnel through the same system means that we can have many sources of pleasure and we can create new ones all the time. This means that we can and do get pleasure from abstract objects like numbers and money that are removed from our basic appetites for food and sex. As long as an object taps into this deep brain system, we find it pleasurable.

When I looked at the Miró prints, I found myself both liking them and wanting one of them in my home. These two different ways of enjoying Miró are rooted in liking and wanting systems of the brain. The typical coupling of liking and wanting makes sense because wanting makes us act to acquire the objects of our desires. Once we acquire these objects, we can get pleasure from them. But the two brain systems can get uncoupled. The distinction between liking and wanting offers a biological interpretation of eighteenth-century ideas advocated by the Earl of Shaftsbury and Immanuel Kant. These thinkers described aesthetic experiences as states of "disinterested interest." If one accepts the framing of aesthetic experiences as disinterested interest (which not everybody does), activating our liking systems without activating the wanting systems is what disinterested interest would mean in the brain.

Even though our pleasures funnel through the same deep brain systems, to say that our pleasure in art is the same as the pleasure we receive from tasting sugar would be silly. Of course aesthetic encounters are more complicated. Aesthetic experiences go far beyond the simple pleasures of our basic appetites. The emotional rewards of aesthetic encounters are more nuanced, and the experiences are more modifiable by our cognitive systems.

A classic example of the nuanced way our emotions can be stirred by aesthetic encounters is again found in the writings of eighteenth-century theoreticians. Edmund Burke explored the beautiful and the sublime in aesthetic experiences. He thought that sublime objects produced a subtle combination of attraction and fear. The grandeur of enormous mountains can be sublime. We experience their awesome beauty at the same time we are faced with our own insignificance. Aesthetic experiences often play with combinations of emotions.

We often experience art as emotional compositions. The observation that the grandeur of a mountain can simultaneously evoke different emotions applies to artwork as well. Contemporary art can evoke complex emotional combinations as it contends with faith, or meditates on obsessive behaviors, or urges us to fight oppressive systems. Art produces awe, fear, passion, fervor, anger, and states of contemplation. As the expressionist theorists of art pointed out, art can communicate nuanced emotions that are hard to convey in words, emotions that can make our hearts race, pupils dilate, and give us chills.

My experience of the contemporary art exhibit, "Love and Death," was very different from my experience of looking at the Picasso and Miró pieces. The baggage I brought to viewing Picasso and Miró was also very different from what I brought to the "Love and Death" exhibit. I have seen many examples of Picasso and Miró paintings and prints. I have read analyses of their works and accounts about their lives. My delight when looking at their art cannot be separated from the knowledge that informed my gaze. For the "Love and Death" art, I knew nothing about the artists, the context in which they produced their work, or what they were trying to accomplish. Earlier, I mentioned that at the core of aesthetic experiences is sensations, emotions, and meaning. I can enjoy Australian aboriginal art because its colors and forms please me, but I do not understand what the art means. Similarly, I can enjoy a Dogon mask, because of its spare form and stylized expression, but I do not know anything about how these masks were used. In these cases, my sensory pleasure is enough for me to have an aesthetic experience. When I don't even have familiar sensations or emotions on which to hang my engagement, as in my first look at the "Love and Death" exhibit, without added information about the work I am completely lost.

Meaning influences even simple pleasures. Knowing the label of a Cola drink influences our enjoyment of its taste. Thinking that an image is generated by a computer or reproduced from a museum makes the reward system in our brain respond differently. What we "see" is just the tip of an aesthetic iceberg. Earlier, I used the example of looking at the text of *Scheherazade* in Arabic calligraphy. I can enjoy the beauty of the visual letter forms without knowing the meaning conveyed by the text. However, if I knew Arabic, my experience of the calligraphy would change dramatically. Similarly, when we can read an artwork, our aesthetic encounter changes dramatically. Looking at the laminated descriptions of the "Love and Death" exhibit allowed me to experience the art differently. This changed experience does not mean that the interpretation of the art on the laminated sheets is the correct one. Rather, the interpretation gave me an introduction through which I was better able to engage with the art.

We encounter limits of what neuroscience can contribute to aesthetics when we consider meaning in art. Neuroscience has something to say about the way we recognize representational paintings. We know something about how we recognize objects or places or faces. In so far as art depicts objects or places or faces we know something about how the brain responds to them. But this knowledge is about our general understanding of these categories of objects and not about the particular response to a Cezanne still life, or a Rembrandt portrait, or a Turner landscape. If you think that the critical level of analysis for aesthetic encounters is the meaning of individual works of art, the way it responds to its place in history, embedded in its local culture, then art's inherent openness to many interpretations is an intractable problem for neuroscience. Understanding the layered meaning of an individual work falls outside the acuity of scientific methods. These methods are best at extracting generalizations. Scientific aesthetics can scrutinize general effects of knowledge on aesthetic encounters, but not the specific knowledge and layered meanings woven into individual works of art.

Neuroaesthetics studies show us that our brains do not have a dedicated aesthetic or art module in the brain. We have no specific aesthetic receptor analogous to our receptors for vision or touch or smell. We have no specific aesthetic emotion analogous to our emotions of fear or anxiety or happiness. We have no specific aesthetic cognition analogous to systems like memory or language or action. Rather, aesthetic experiences flexibly engage neural ensembles of sensory, emotional, and cognitive systems. This flexibility built into the ensembles is part of what makes art and aesthetic experiences varied and unpredictable.

We do not have an art instinct. I suspect that this pronouncement will not sit well with many. The idea of art as an instinct is reassuring because it implies that art is really important; that art is a deep and integral part of our being; that art is a universal human preoccupation. People who denigrate art in education, public policy, and communal discourse are denigrating human nature. A deep worry for art lovers is that if art does not reflect an instinct, then we might conclude that art is trivial, frivolous, and a luxury born of an indulgent society. By now, it should be clear that such a conclusion is not warranted. As I mentioned before, we do not have an instinct for reading and writing, yet, few would argue that reading and writing is trivial, frivolous, or a luxury born of an indulgent society.

Art is everywhere and has existed in some form as far back as we can tell. The universality of art makes it unlikely that art is simply a by-product of other evolved cognitive capacities. Scholars typically frame art as an instinct or as an evolutionary by-product. We need a third way to think of art that accepts both its universality bred in our brains and its variety fashioned by history. In the last chapter, I used the example of the white rumped munia's and Bengalese finch's song to suggest a third way that acknowledges art's instinctual roots and welcomes its cultural blossoming. The relaxation of instinctual constraints on behavior is precisely what allows art to evolve flexibly and to be so wonderfully surprising.

By emphasizing art as a response to local conditions, are we missing something important? After all, doesn't the most inspiring art engage themes that we all face, have always faced, and will always face? The exhibit that I found bewildering was called "Love and Death." Are there any more universal themes? My response to these questions is that even when art addresses universal themes, it does so in a local idiom. Just because love and death are universal themes did not mean that I could engage with scattered little birds beneath the branches of an upside-down tree. The power of art, its ability to move us and make us experience old themes with new eyes, is conveyed through its local expression. The content of art is shaped by local conditions: the culture in which it is born, its historical antecedents, the economic conditions of its production and reception, and references relevant to its time and place.

If beauty is a mongrel, then art is a chimera. Art is a messy collection of adaptations, spandrels, and exaptations, and it is replete with modifications and plug-ins fashioned by historical episodes and cultural niches. When cultural pressures select specific kinds of art, the art produced falls within narrow stylized boundaries. When cultural selective pressures are relaxed, art blossoms. We do not have a single art instinct. We have instincts that trigger art-like behavior. Rather than being dominated by instincts, it is the relaxation of instinctual control that allows art to express itself fully. Art germinates instinctually and matures serendipitously. Its content is a serendipitous mixture born of time and place and culture and personality. Could it be any other way? Being deprived of a grand unifying instinctual theory of art is not cause for concern. Instead, the diverse, local, and serendipitous nature of art is precisely why art can surprise us, enlighten us, force us to see the world differently, ground us, shake us, please us, anger us, bewilder us, and make believers of us.

When free, we relax into art. We are better off for it.

ACKNOWLEDGMENTS

Like laws and sausage, the making of a book can be ugly. One feels the burden of this ugliness all the more when writing about aesthetics. I have been fortunate to have many supporters who encouraged and helped me along the way to make this book less ugly. They have my deep gratitude and deserve special mention.

Lisa Santer, my partner for over 25 years, has read most of my professional writing. She had the privilege of reading this manuscript at its ugliest. My most critical editor, she spent many hours reading the first draft and exhorting me to be clear about what I wanted to say. If you read the book, you should also be grateful for her scrubbing of the first draft.

My hope is that the book will appeal to scholars in the humanities and the sciences and the interested general reader. Again, I was fortunate to have input from many smart people with knowledge, expertise, and just plain good sense that made this book better. These people are extremely busy with their own professional and personal lives. I am grateful that they took time to read my book even as I was stumbling through my ideas.

Marcos Nadal is a psychologist who has deep knowledge of human evolution and empirical aesthetics. His invitation to visit Mallorca in the fall of 2010 set this project in motion. His careful reading and direction gave this book a texture that would otherwise not have been possible.

Oshin Vartanian is a psychologist and cognitive neuroscientist who has been engaged with experimental neuroaesthetics from the beginning. His knowledge of psychology and empirical aesthetics was extremely helpful in fitting my text within a broader empirical narrative.

Jonna Kwaitkowski is a psychologist who has wide-ranging knowledge of empirical aesthetics. Her enthusiasm for the project buoyed me over the months, and her insightful intuitions about "the voice" I should adopt set the tone for much of the book.

Helmut Leder is a cognitive psychologist who probably knows as much about empirical aesthetics as anyone in the world as of this writing. A marvelous host for my trip to Vienna in 2011 while I was writing this book, I appreciate his input on an early draft.

Russell Epstein is a cognitive neuroscientist and colleague who was a check on whether I was simplifying the neuroscience too much and whether my delving into the humanities was comprehensible. I appreciate his good sense and his suggestions to repeat my key points so that readers would not get too lost in my verbal meanderings.

William Seeley is philosopher with an abiding interest in neuroaesthetics. His feedback was invaluable. Like any good philosopher, he pointed out where my thinking was muddled, especially in the Art section.

Noël Carroll is a senior philosopher who has deep interest in and knowledge of aesthetics and the philosophy of art. His wise counsel was extremely helpful. Taking his targeted suggestions into consideration vastly improved the final version of the book. His enthusiasm for the project was deeply reassuring.

Blake Gopnik is an art critic who is fundamentally skeptical of scientific aesthetics. It is a luxury to have a reader who disagrees with your enterprise and is willing to spend the time to explain why. His trenchant comments were invaluable in shaping this book.

Raphael Rosenberg is an art historian who has an interest in empirical methods. I met him briefly in Vienna. Based on this brief meeting, he was willing to read the book and offer a perspective that broadened my view. His enthusiasm and support for my approach was heartening.

Joseph Kable is a cognitive neuroscientist, former student, and current friend who is expert on human decision-making and reward systems. He read the Pleasure section and was my guide to whether I got that part right in a way that would not make a card-carrying neuroeconomist nauseous.

Daniel Graham is a mathematical psychologist who probably understood immediately why Euler's identity is beautiful. He read the chapters on numbers and on experimental science of the arts and reassured me that I made sense.

Jennifer Murphy, my dear friend, helped me target the book for its intended non-specialist reader. An English major in college, she was sensitive to my use of language. Working in science now, but not neuroscience or psychology or evolutionary biology, she noticed the bumps in my exposition that needed smoothing. Her ongoing support was invaluable.

One nice thing about working at a place like the University of Pennsylvania is the wide and deep expertise of people around. I want to acknowledge Robert Kurzban and Marc Schmidt. Kurzban is a bona fide evolutionary psychologist and Schmidt studies the neural bases of bird songs. They both took time to have lunch with me and let me bounce off my ideas on evolutionary psychology and bird songs, respectively. They allayed some of my fears about incorporating ideas from their fields (in which I claim no special expertise) into this book.

Finally, I am grateful for the support of the staff at Oxford University Press. The production editor, Emily Perry, and the copy editor, Jerri Hurlbutt, were excellent. The assistant editor, Miles Osgood, and marketing manager, John Hercel, were assiduous in their communications. I wish to acknowledge the skills of the design team at Oxford. Their graphic representation of the ventral brain for the cover of this book delighted me. Catherine Alexander, my first editor, was instrumental in my starting this project. She believed that such a book would be timely and was confident in my ability to write it. After she left Oxford, Joan Bossert took over ably. Her patience, advice, and enthusiastic encouragement resulted in the book before you.

BOOKS (THAT I FOUND USEFUL)

Barkow, J. H., Cosmides, L., & Tooby, J. (Eds.). (1992). *The Adapted Mind: Evolutionary Psychology and the Generation of Culture*. New York: Oxford University Press.

Buller, D. J. (2005). *Adapting Minds: Evolutionary Psychology and the Persistent Quest for Human Nature*. Cambridge, MA: MIT Press.

Burke, E. (1757/1998). *A Philosophical Inquiry into the Origin of Our Ideas of the Sublime and Beautiful*. New York: Oxford University Press.

Butler, C. (2004). *Pleasure and the Arts: Enjoying Literature, Painting, and Music*. New York: Oxford University Press.

Carroll, N. (2010). *Art in Three Dimensions*. New York: Oxford University Press.

Carroll, N. (Ed.). (2000). *Theories of Art Today*. Madison, WI: University of Wisconsin Press.

Clark, K. (1956). *The Nude: A Study in Ideal Form*. Princeton, NJ: Princeton University Press.

Collins, M. (1999). *This is Modern Art*. London: Weidenfield & Nicolson.

Coyne, J. (2009). *Why Evolution is True*. New York: Viking.

Curtis, G. (2007). *The Cave Painters: Probing the Mysteries of the World's First Artists*. New York: Anchor Books.

Davies, S. (2006). *The Philosophy of Art*. Malden, MA: Blackwell Publishing.

Dissanayake, E. (1988). *What is Art For?* Seattle: University of Washington Press.

Dutton, D. (2009). *The Art Instinct. Beauty, Pleasure, and Human Evolution*. New York: Bloomsbury Press.

Etcoff, N. (1999). *Survival of the Prettiest*. New York: Anchor Books.

Feagin, S. L., & Maynard, M. (Eds.). (1997). *Aesthetics*. New York: Oxford University Press.

Freeland, C. (2001). *But Is it Art? An Introduction to Art Theory*. New York: Oxford University Press.

Gombrich, E. (1960). *Art and Illusion*. Princeton, NJ: Princeton University Press.

Kringlebach, M., & Berridge, K. C. (Eds.). (2009). *Pleasures of the Brain*. New York: Oxford University Press.

Livingstone, M. (2002). *Vision and Art: The Biology of Seeing*. New York: Abrams.

Livio, M. (2002). *The Golden Ratio. The Story of Phi, the World's Most Astonishing Number*. New York: Broadway Books.

Mayr, E. (2001). *What Evolution Is*. New York: Basic Books.

Miller, G. (2000). *The Mating Mind: How Sexual Choice Shaped the Evolution of Human Nature*. New York: Doubleday.

Ogas, O., & Gaddam, S. (2011). *A Billion Wicked Thoughts. What the World's Largest Experiment Reveals About Human Desire*. New York: Dutton.

Onians, J. (2008). *Neuroarthistory: From Aristotle and Pliney to Baxandall and Zeki*. New Haven, CT: Yale University Press.

Pallen, M. (2009). *The Rough Guide to Evolution*. London: Rough Guides.

Pinker, S. (1997). *How the Mind Works*. New York: W.W. Norton.

Rhodes, G., & Zebrowitz, A. (Eds.). (2002). *Facial Attractiveness. Evolutionary, Cognitive, and Social Perspectives*. Westport, CT: Ablex Publishing.

Roach, M. (2008). *Bonk: The Curious Coupling of Science and Sex*. New York: W.W. Norton.

Santayana, G. (1896). *The Sense of Beauty. Being the Outline of Aesthetic Theory*. New York: Dover Publications.

Sartwell, C. (2004). *Six Names of Beauty*. New York: Routledge.

Scruton, R. (2009). *Beauty*. New York: Oxford University Press.

Shermer, M. (2008). *The Mind of the Market: How Biology and Psychology Shape Our Economic Lives*. New York: Holt.

Shimamura, A. P., & Palmer, S. E. (Eds.). (2012). *Aesthetic Science: Connecting Minds, Brains and Experience*. New York: Oxford University Press.

Shiner, L. (2001). *The Invention of Art. A Cultural History*. Chicago: University of Chicago Press.

Skov, M., & Vartanian, O. (Eds.). (2009). *Neuroaesthetics*. Amityville, NY: Baywood Publishing.

Turner, M. (Ed.). (2006). *The Artful Mind. Cognitive Science and the Riddle of Human Creativity*. New York: Oxford University Press.

Wallenstein, G. W. (2009). *The Pleasure Instinct: Why We Crave Adventure, Chocolate, Pheromones, and Music*. Hoboken, NJ: John Wiley & Sons.

Weaver, J. H. (2003). *The Math Explorer: A Journey Through the Beauty of Mathematics*. Amherst., NY: Prometheus Books.

Weintraub, L., Danto, A., & McEvilley, T. (Eds.). (1996). *Art on the Edge and Over*. Litchfield, CT: Art Insights.

Zeki, S. (1999). *Inner Vision: An Exploration of Art and the Brain*. New York: Oxford University Press.

REFERENCES

1. Shimamura, A. P. Towards a science of aesthetics. In *Aesthetic Science: Connecting Minds, Brains and Experience*, A. P. Shimamura & S. E. Palmer (Eds). 2012, New York: Oxford University Press, pp. 3–28.
2. *Most Beautiful Women of the 20th century*. 2010 Accessed December 2, 2012, from http://movies.rediff.com/slide-show/2010/jul/01/slide-show-1-most-beautiful-women.htm.
3. Cunningham, M. R., et al. "Their ideas of beauty are, on the whole, the same as ours": Consistency and variability in the cross-cultural perception of female physical attractiveness. *Journal of Personality and Social Psychology*, 1995. *68*(2): p. 261.
4. Langlois, J. H., et al. Maxims or myths of beauty? A meta-analytic and theoretical review. *Psychological Bulletin*, 2000. *126*(3): pp. 390–423.
5. Jones, D., & Hill, K. Criteria of facial attractiveness in five populations. *Human Nature*, 1993. *4*(3): pp. 271–296.
6. Mondloch, C. J., et al. Face perception during early infancy. *Psychological Science*, 1999. *10*(5): pp. 419–422.
7. Pascalis, O., et al. Mother's face recognition by neonates: A replication and an extension. *Infant Behavior and Development*, 1995. *18*(1): pp. 79–85.
8. Slater, A., et al. Newborn infants prefer attractive faces. *Infant Behavior and Development*, 1998. *21*(2): pp. 345–354.
9. Langlois, J. H., et al. Facial diversity and infant preferences for attractive faces. *Developmental Psychology*, 1991. *27*(1): pp. 79–84.
10. Langlois, J. H., Roggman, L. A., & Rieser-Danner, L.A. Infants' differential social responses to attractive and unattractive faces. *Developmental Psychology*, 1990. *26*(1): p. 153.

11. Leder, H., et al. When attractiveness demands longer looks: The effects of situation and gender. *Quarterly Journal of Experimental Psychology,* 2010. *63*(9): pp. 1858–1871.
12. Bergman, J. Using facial angle to prove evolution and the human race hierarchy. *Journal of Creation,* 2010. *24*: pp. 101–105.
13. Lavater, J. C. *Essays on Physiognomy.* 15th ed. 1878, London: William Tegg and Co.
14. Blackford, K. M. H., & Newcomb, A. *The Job, the Man, the Boss.* 1919, Garden City, NY: Doubleday.
15. Galton, F. Composite portraits, made by combining those of many different persons into a single resultant figure. *Journal of the Anthropological Institute of Great Britain and Ireland,* 1878. *8*: pp. 132–142.
16. Thornhill, R., & Gangestad, S. W. Facial attractiveness. *Trends in Cognitive Sciences,* 1999. *3*(12): pp. 452–260.
17. Rubenstein, A. J., Kalakanis, L., & Langlois, J. H. Infant preferences for attractive faces: A cognitive explanation. *Developmental Psychology,* 1999. *35*(3): pp. 848–855.
18. Grammer, K., & Thornhill, R. Human (*Homo sapiens*) facial attractiveness and sexual selection: The role of symmetry and averageness. *Journal of Comparative Psychology,* 1994. *108*(3): pp. 233–242.
19. Mealey, L., Bridgstock, R., & Townsend, G. C. Symmetry and perceived facial attractiveness: A monozygotic co-twin comparison. *Journal of Personality and Social Psychology,* 1999. *76*(1): pp. 151–158.
20. Buss, D. M. Sex differences in human mate preferences: Evolutionary hypotheses tested in 37 cultures. *Behavioral and Brain Sciences,* 1989. *12*(01): pp. 1–14.
21. Grammer, K., et al. Darwinian aesthetics: Sexual selection and the biology of beauty. *Biological Reviews,* 2003. *78*(3): pp. 385–407.
22. Perrett, D. I., May, K. A., & Yoshikawa, S. Facial shape and judgements of female attractiveness. *Nature,* 1994. *368*: pp. 239–242.
23. Winkler, E. M., & Kirchengast, S. Body dimensions and differential fertility in Kung San males from Namibia. *American Journal of Human Biology,* 1994. *6*(2): pp. 203–213.
24. Mazur, A., Mazur, J., & Keating, C. Military rank attainment of a West Point class: Effects of cadets' physical features. *American Journal of Sociology,* 1984. *90*(1): pp. 125–150.
25. Perrett, D. I., et al. Effects of sexual dimorphism on facial attractiveness. *Nature,* 1998. *394*: pp. 884–887.
26. Penton-Voak, I. S., & Perrett, D. I. Female preference for male faces changes cyclically: Further evidence. *Evolution and Human Behavior,* 2000. *21*(1): pp. 39–48.
27. Møller, A. P., & Thornhill, R. Bilateral symmetry and sexual selection: A meta-analysis. *American Naturalist,* 1998. *151*(2): pp. 174–192.

28. Markusson, E., & Folstad, I. Reindeer antlers: Visual indicators of individual quality? *Oecologia*, 1997. *110*(4): pp. 501–507.
29. Moller, A. P. Female swallow preference for symmetrical male sexual ornaments. *Nature*, 1992. *357*(6375): pp. 238–240.
30. Thornhill, R., & Gangestad, S. W. Human fluctuating asymmetry and sexual behavior. *Psychological Science*, 1994. *5*(5): pp. 297–302.
31. Thornhill, R., Gangestad, S. W., & Comer, R. Human female orgasm and mate fluctuating asymmetry. *Animal Behaviour*, 1995. *50*(6): pp. 1601–1615.
32. Scutt, D., & Manning, J. T. Ovary and ovulation: Symmetry and ovulation in women. *Human Reproduction*, 1996. *11*(11): pp. 2477–2480.
33. Jackson, L. A., & Ervin, K. S. Height stereotypes of women and men: The liabilities of shortness for both sexes. *Journal of Social Psychology*, 1992. *132*(4): pp. 433–445.
34. Zeifman, D. M., & Ma, J. E. Experimental examination of women's selection criteria for sperm donors versus life partners. *Personal Relationships*, 2012. doi: 10.1111/j.1475-6811.2012.01409.x
35. Horvath, T. Correlates of physical beauty in men and women. *Social Behavior and Personality*, 1979. *7*(2): pp. 145–151.
36. Pettijohn, T. F., & Jungeberg, B. J. Playboy Playmate curves: Changes in facial and body feature preferences across social and economic conditions. *Personality and Social Psychology Bulletin*, 2004. *30*(9): pp. 1186–1197.
37. Pettijohn Ii, T. F., & Tesser, A. Popularity in environmental context: Facial feature assessment of American movie actresses. *Media Psychology*, 1999. *1*(3): pp. 229–247.
38. Singh, D. Adaptive significance of female physical attractiveness: Role of waist-to-hip ratio. *Journal of Personality and Social Psychology*, 1993. *65*(2): pp. 293–307.
39. Confer, J. C., Perilloux, C., & Buss, D. M. More than just a pretty face: Men's priority shifts toward bodily attractiveness in short-term versus long-term mating contexts. *Evolution and Human Behavior*, 2010. *31*(5): pp. 348–353.
40. Sivinski, J., & Burk, T. Reproductive and mating behaviour. In *Fruit Flies: Their Biology, Natural Enemies, and Control*, A. S. Robinson & G. Hooper (Eds). 1989. New York: Elsevier, p. 343.
41. Singer, F., et al. Analysis of courtship success in the funnel-web spider *Agelenopsis aperta*. *Behaviour*, 2000. *137*(1): pp. 93–117.
42. Manning, J. T., & Pickup, L. J. Symmetry and performance in middle distance runners. *International Journal of Sports Medicine*, 1998. *19*(03): pp. 205–209.
43. Provost, M. P., Troje,N. F., & Quinsey, V. L. Short-term mating strategies and attraction to masculinity in point-light walkers. *Evolution and Human Behavior*, 2008. *29*(1): pp. 65–69.
44. Kenealy, P., Frude, N., & Shaw, W. Influence of children's physical attractiveness on teacher expectations. *Journal of Social Psychology*, 1988. *128*: pp. 373–383.

45. Sroufe, R., et al. The effects of physical attractiveness on honesty: A socially desirable response. *Personality and Social Psychology Bulletin*, 1976. *3*(1): pp. 59–62.
46. Benson, P. L., Karabenick, S. A., & Lerner, R. M. Pretty pleases: The effects of physical attractiveness, race, and sex on receiving help. *Journal of Experimental Social Psychology*, 1976. *12*(5): pp. 409–415.
47. Chatterjee, A., et al. The neural response to facial attractiveness. *Neuropsychology*, 2009. *23*(2): pp. 135–143.
48. Aharon, I., et al. Beautiful faces have variable reward value: fMRI and behavioral evidence. *Neuron*, 2001. *32*: pp. 537–551.
49. Winston, J., et al., Brain systems for assessing facial attractiveness. *Neuropsychologia*, 2007. *45*: pp. 195–206.
50. Kim, H., et al. Temporal isolation of neural processes underlying face preference decisions. *Proceedings of the National Academy of Sciences U S A*, 2007. *104*(46): pp. 18253–18258.
51. Brown, S., Martinez, M. J., & Parsons, L. M. The neural basis of human dance. *Cerebral Cortex*, 2006. *16*(8): pp. 1157–1167.
52. Calvo-Merino, B., et al. Towards a sensorimotor aesthetics of performing art. *Consciousness and Cognition*, 2008. *17*(3): pp. 911–922.
53. Miller, G. F. Sexual selection for cultural displays. In *The Evolution of Culture*, R. Dunbar, C. Knight, & C. Power (Eds). 1999, New Brunswick, NJ: Rutgers University Press, pp. 71–91.
54. Mervis, C. B., & Rosch, E. Categorization of natural objects. *Annual Review of Psychology*, 1981. *32*(1): pp. 89–115.
55. Martindale, C., & Moore, K. Priming, prototypicality, and preference. *Journal of Experimental Psychology: Human Perception and Performance*, 1988. *14*: pp. 661–667.
56. Gangestad, S. W., & Buss, D. M. Pathogen prevalence and human mate preferences. *Ethology and Sociobiology*, 1993. *14*(2): pp. 89–96.
57. Ruff, C. B., & Jones, H. H. Bilateral symmetry in cortocal bone of the humerus and tibia—sex and age factors. *Human Biology*, 1981. *53*: pp. 69–86.
58. Grammer, K., et al. Female faces and bodies: n-dimensional feature space and attractiveness. In *Facial Attractiveness: Evolutionary, Cognitive, and Social Perspectives*, G. Rhodes & L. Zebrowitz (Eds.). 2001, Westport, CT: Ablex.
59. Zahavi, A., & Zahavi, A. *The Handicap Principle: A Missing Piece of Darwin's Puzzle*. 1997, Oxford: Oxford University Press.
60. Hamilton, W. D., & Zuk, M. Heritable true fitness and bright birds: A role for parasites? *Science*, 1982. *218*(4570): pp. 384–387.
61. Little, A. C., DeBruine, L. M., & Jones, B. C. Exposure to visual cues of pathogen contagion changes preferences for masculinity and symmetry in opposite-sex faces. *Proceedings of the Royal Society, Series B: Biological Sciences*, 2011. *278*(1714): pp. 2032–2039.

62. Perrett, D. I., et al. Effects of sexual dimorphism on facial attractiveness. *Nature,* 1998. *394*(6696): pp. 884–887.
63. Gangestad, S. W., Thornhill, R., & Garver-Apgar, C. E. Adaptations to ovulation implications for sexual and social behavior *Current Directions in Psychological Science,* 2005. *14*(6): pp. 312–316.
64. Apicella, C. L., & Feinberg, D. R. Voice pitch alters mate-choice-relevant perception in hunter-gatherers. *Proceedings of the The Royal Society: Biological Sciences/,* 2009. *276*(1659): pp. 1077–1082.
65. Møller, A. P. Ejaculate quality, testes size and sperm competition in primates. *Journal of Human Evolution,* 1988. *17*(5): pp. 479–488.
66. Tinbergen, N. *Curious Naturalist* 1954, New York: Basic Books.
67. Ramachandran, V. S., & Hirstein, H. The science of art: A neurological theory of aesthetic experience. *Journal of Consciousness Studies,* 1999. *6*: pp. 15–51.
68. Daprati, E., Iosa, M., & Haggard, P. A dance to the music of time: Aesthetically relevant changes in body posture in performing art. *PLoS ONE,* 2009. *4*(3): p. e5023.
69. Corson, R. *Fashions in Makeup, from Ancient to Modern Times* 1972, Owen.
70. Knight, C., Power, C., & Watts, I. The human symbolic revolution: A Darwinian account. *Cambridge Archaeological Journal,* 1995. *5*(1): pp. 75–114.
71. Gallup, J. G. G., & Frederick, D. A. The science of sex appeal: An evolutionary perspective. *Review of General Psychology,* 2010. *14*(3): p. 240.
72. Kaplan, S.,. Kaplan, R., & Wendt, J. Rated preference and complexity for natural and urban visual material. *Perception & Psychophysics,* 1972. *12*(4): pp. 354–356.
73. Balling, J. D., & Falk, J. H. Development of visual preference for natural environments. *Environment and Behavior,* 1982. *14*(1): pp. 5–28.
74. Synek, E., & Grammer. K. *Evolutionary Aesthetics: Visual Complexity and the Development of Landscape Preference.* 1998 Retrieved November 21, 2012, from http://evolution.anthro.univie.ac.at/institutes/urbanethology/projects/urbanisation/landscapes/indexland.html.
75. Heerwagen, J. H., & Orians, G. H. Humans, habitats, and aesthetics. In *The Biophilia Hypothesis* S. R. Kellert & E. O. Wilson (Eds.). 1995, Washinton, DC: Island Press.
76. Han, K. T. An exploration of relationships among the responses to natural scenes: Scenic beauty, preference, and restoration. *Environment and Behavior,* 2010. *42*(2): pp. 243–270.
77. Orians, G. H., & Heerwagen, J. H. Evolved responses to landscapes. In *The Adapted Mind: Evolutionary Psychology and the Generation of Culture,* J. H. Barkow, L. Cosmides, & J. Tooby (Eds.). 1992, : New York: Oxford University Press, pp. 555–580.
78. Kaplan, R., & Kaplan, S. *The Experience of Nature: A Psychological Perspective* 1989, New York: Cambridge University Press.

79. Epstein, R. A., & Higgins, J. S. Differential parahippocampal and retrosplenial involvement in three types of visual scene recognition. *Cerebral Cortex,* 2006. *17*(7): pp. 1680–1693.
80. Yue, X., Vessel, E. A., & Biederman, I. The neural basis of scene preferences. *Neuroreport,* 2007. *18*(6): pp. 525–529.
81. Mitchison, G. J. Phyllotaxis and the Fibonacci series. *Science,* New Series, 1977. *196*(4287): pp. 270–275.
82. Douady, S., & Couder, Y. Phyllotaxis as a physical self-organized growth process. *Physical Review Letters,* 1992. *68*(13): pp. 2098–2101.
83. Schmidhuber, J. Low-complexity art. *Leonardo,* 1997. *30*(2): pp. 97–103.
84. Gerstmann, J., Some notes on the Gerstmann syndrome. *Neurology,* 1957. 7(12): pp. 866–866.
85. Dehaene, S., et al. Three parietal circuits for number processing. *Cognitive Neuropsychology,* 2003. *20*(3–6): p. 487–506.
86. Schmidt, H. J., & Beauchamp, G. K. Adult-like odor preferences and aversions in three-year-old children. *Child Development,* 1988. *59*(4): pp. 1136–1143.
87. Liley, A. W. The foetus as a personality. *Australian and New Zealand Journal of Psychiatry,* 1972. *6*(2): pp. 99–105.
88. Faas, A.E., et al. Differential responsiveness to alcohol odor in human neonates: Efects of maternal consumption during gestation. *Alcohol,* 2000. *22*(1): pp. 7–17.
89. Schaal, B., Marlier, L., & Soussignan, R. Human foetuses learn odours from their pregnant mother's diet. *Chemical Senses,* 2000. *25*(6): pp. 729–737.
90. Gottfried, J. A., Winston, J. S., & Dolan, R. J. Dissociable codes of odor quality and odorant structure in human piriform cortex. *Neuron,* 2006. *49*(3): pp. 467–479.
91. Rolls, E. T., Kringelbach, M. L., & de Araujo, I. E. T. Different representations of pleasant and unpleasant odours in the human brain. *European Journal of Neuroscience,* 2003. *18*(3): pp. 695–703.
92. Anderson, A. K., et al. Dissociated neural representations of intensity and valence in human olfaction. *Nature Neuroscience,* 2003. *6*(2): pp. 196–202.
93. Small, D. M., et al., Differential neural responses evoked by orthonasal versus retronasal odorant perception in humans. *Neuron,* 2005. *47*(4): pp. 593–605.
94. Veldhuizen, M. G., Rudenga, K. J., & Small, D. M. The pleasure of taste, flavor, and food. In *Pleasures of the Brain,* M. Kringlebach & K. C. Berridge (Eds.). 2009, New York: Oxford University Press, pp. 146–168.
95. O'Doherty, J., et al. Sensory-specific satiety-related olfactory activation of the human orbitofrontal cortex. *Neuroreport,* 2000. *11*(4): pp. 893–897.
96. Kringelbach, M. L., et al. Activation of the human orbitofrontal cortex to a liquid food stimulus is correlated with its subjective pleasantness. *Cerebral Cortex,* 2003. *13*(10): pp. 1064–1071.
97. Gottfried, J. A., O'Doherty, J., & Dolan, R. J. Appetitive and aversive olfactory learning in humans studied using event-related functional magnetic resonance imaging. *Journal of Neuroscience,* 2002. *22*(24): pp. 10829–10837.

98. Schultz, W., Dayan, P., & Montague, P. R. A neural substrate of prediction and reward. *Science*, 1997. *275*: pp. 1593–1598.
99. O'Doherty, J. P., et al. Temporal difference models and reward-related learning in the human brain. *Neuron*, 2003. *38*: pp. 329–337.
100. McClure, S. M., et al. Neural correlates of behavioral preference for culturally familiar drinks. *Neuron*, 2004. *44*(2): pp. 379–387.
101. de Araujo, I. E., et al. Cognitive modulation of olfactory processing. *Neuron*, 2005. *46*: pp. 671–679.
102. Plassmann, H., et al. Marketing actions can modulate neural representations of experienced pleasantness. *Proceedings of the National Academy of Sciences U S A*, 2008. *105*: pp. 1050–1054.
103. Wansink, B., Payne, C. R., & North, J. Fine as North Dakota wine: Sensory expectations and the intake of companion foods. *Physiology & Behavior*, 2007. *90*(5): pp. 712–716.
104. Hurst, W. J., et al. Cacao usage by the earliest Maya civilization. *Nature*, 2002. *418*(6895): pp. 289–290.
105. Martin, F.-P. J., et al. Metabolic effects of dark chocolate consumption on energy, gut microbiota, and stress-related metabolism in free-living subjects. *Journal of Proteome Research*, 2009. *8*(12): pp. 5568–5579.
106. Barr, R. G., et al. Effects of intra-oral sucrose on crying, mouthing and hand-mouth contact in newborn and six-week-old infants. *Developmental Medicine & Child Neurology*, 1994. *36*(7): pp. 608–618.
107. di Tomaso, E., Beltramo, M., & Piomelli, D. Brain cannabinoids in chocolate. *Nature*, 1996. *382*(6593): pp. 677–678.
108. Crawford, M. A., et al. Evidence for the unique function of docosahexaenoic acid during the evolution of the modern hominid brain. *Lipids*, 1999. *34*: pp. 39–47.
109. Drewnowski, A., & Greenwood, M. Cream and sugar: Human preferences for high-fat foods. *Physiology & Behavior*, 1983. *30*(4): pp. 629–633.
110. Pittman, D. W., et al. Linoleic and oleic acids alter the licking responses to sweet, salt, sour, and bitter tastants in rats. *Chemical Senses*, 2006. *31*(9): pp. 835–843.
111. Volkow, N. D., et al. Overlapping neuronal circuits in addiction and obesity: Evidence of systems pathology. *Philosophical Transactions of the Royal Society. Series B: Biological Sciences*, 2008. *363*(1507): pp. 3191–3200.
112. Klüver, H., & Bucy, P. C. "Psychic blindness" and other symptoms following bilateral temporal lobectomy in Rhesus monkeys. *American Journal of Physiology*, 1937. *119*: pp. 352–353.
113. Lilly, R., et al. The human Klüver-Bucy syndrome. *Neurology*, 1983. 33(9): pp. 1141–1141.
114. Laumann, E. O., et al. The social organization of sexuality: Sexual practices in the United States. In *Studies in Crime and Justice Series* 1994, Chicago: University of Chicago Press.

115. Cameron, P., & Biber, H. Sexual thought throughout the life-span. *Gerontologist*, 1973. *13*(2): pp. 144–147.
116. *Pornography industry is larger than the revenues of the top technology*. 2010. Retrieved November 24, 2012, from http://blog.cytalk.com/2010/01/web-porn-revenue/.
117. Deaner, R. O., Khera, A. V., & Platt, M. L. Monkeys pay per view: Adaptive valuation of social images by rhesus macaques. *Current Biology*, 2005. *15*(6): pp. 543–548.
118. Hamann, S., et al. Men and women differ in amygdala response to visual sexual stimuli. *Nature Neuroscience*, 2004. *7*(4): p. 411–416.
119. Lal, S., et al. Apomorphine: Clinical studies on erectile impotence and yawning. *Progress in Neuro-Psychopharmacology and Biological Psychiatry*, 1989. *13*(3–4): pp. 329–339.
120. Whipple, B., & Komisaruk, B. R. Elevation of pain threshold by vaginal stimulation in women. *Pain*, 1985. *21*(4): pp. 357–367.
121. Phillips, A. G., Vacca, G., & Ahn, S. A top-down perspective on dopamine, motivation and memory. *Pharmacology Biochemistry and Behavior*, 2008. *90*(2): pp. 236–249.
122. Arnow, B. A., et al. Brain activation and sexual arousal in healthy, heterosexual males. *Brain*, 2002. *125*(5): pp. 1014–1023.
123. Georgiadis, J. R., & Kotekaas, R. The sweetest taboo: Funtional neurobiology of human sexuality in relation to pleasure. In *Pleasures of the Brain*, M. Kringlebach & K. C. Berridge (Eds.). 2009, New York: Oxford University Press, pp. 178–201.
124. Baumgartner, T., et al. Oxytocin shapes the neural circuitry of trust and trust adaptation in humans. *Neuron*, 2008. *58*(4): pp. 639–650.
125. Davis, K. D. The neural circuitry of pain as explored with functional MRI. *Neurological Research*, 2000. *22*(3): pp. 313–317.
126. Rachman, S., & Hodgson, R. J. Experimentally induced "sexual fetishism": Replication and development. *Psychological Record*, 1968. *18*(1): pp. 25–27.
127. Moan, C. E., & Heath, R. G. Septal stimulation for the initiation of heterosexual behavior in a homosexual male. *Journal of Behavior Therapy and Experimental Psychiatry*, 1972. *3*(1): pp. 23–30.
128. Broca, P. M. Remarques sur le siége de la faculté du langage articulé,suivies d'une observation d'aphémie (perte de la parole). *Bulletin de la Société Anatomique*, 1861. *6*: pp. 330–357.
129. Pessiglione, M., et al. How the brain translates money into force: A neuroimaging study of subliminal motivation. *Science*, 2007. *316*(5826): pp. 904–906.
130. Knutson, B., et al. Distributed neural representation of expected value. *Journal of Neuroscience.*, 2005, *25*(19): pp. 4806–4812.
131. Seymour, B., et al. Differential encoding of losses and gains in the human striatum. *Journal of Neuroscience*, 2007. *27*: pp. 4826–4831.

132. Sarinopoulos, I., et al. Uncertainty during anticipation modulates neural responses to aversion in human insula and amygdala. *Cerebral Cortex,* 2010. *20*(4): pp. 929–940.
133. Sescousse, G., Redouté, J., & Dreher, J.-C. The architecture of reward value coding in the human orbitofrontal cortex. *Journal of Neuroscience,* 2010. *30*(39): pp. 13095–13104.
134. Tversky, A., & Kahneman, D. The framing of decisions and the psychology of choice. *Science,* 1981. *211*(4481): pp. 453–458.
135. Kahneman, D., Knetsch, J. L., & Thaler, R. H. Anomalies: The endowment effect, loss aversion, and status quo bias. *Journal of Economic Perspectives,* 1991. *5*(1): pp. 193–206.
136. Prelec, D., & Simester, D. Always leave home without it: A further investigation of the credit-card effect on willingness to pay. *Marketing Letters,* 2001. *12*(1): pp. 5–12.
137. Rangel, A., Camerer, C., & Montague, P. R. A framework for studying the neurobiology of value-based decision making. *Nature Reviews Neuroscience,* 2008. *9*(7): pp. 545–556.
138. Seymour, B., & Dolan, R. Emotion, decision making, and the amygdala. *Neuron,* 2008. *58*(5): pp. 662–671.
139. Skinner, B. F. Reinforcement today. *American Psychologist,* 1958. *13*(3): p. 94.
140. Kable, J. W., & Glimcher, P. W. The neural correlates of subjective value during intertemporal choice. *Nature Neuroscience.,* 2007. *10*: pp. 1625–1633.
141. Shiv, B., & Fedorikhin, A. Heart and mind in conflict: The interplay of affect and cognition in consumer decision making. *Journal of Consumer Research,* 1999. *26*(3): pp. 278–292.
142. Mellers, B. A., & McGraw, A. P. Anticipated emotions as guides to choice. *Current Directions in Psychological Science,* 2001. *10*(6): pp. 210–214.
143. Dawkins, R. *The God Delusion* 2006, New York: Bantam Press.
144. Berridge, K. C., Robinson, T. E., & Aldridge, J. W. Dissecting components of reward: "liking," "wanting," and learning. *Current Opinion in Pharmacology,* 2009. *9*: pp. 65–73.
145. Berridge, K., & Kringelbach, M. Affective neuroscience of pleasure: Reward in humans and animals. *Psychopharmacology,* 2008. *199*(3): pp. 457–480.
146. Smith, K. S., & Berridge, K. C. Opioid limbic circuit for reward: Interaction between hedonic hotspots of nucleus accumbens and ventral pallidum. *Journal of Neuroscience,* 2007. *27*(7): pp. 1594–1605.
147. Berridge, K. C. Food reward: Brain substrates of wanting and liking. *Neuroscience & Biobehavioral Reviews,* 1996. *20*(1): pp. 1–25.
148. Schultz, W. Multiple dopamine functions at different time courses. *Annual Review of Neuroscience,* 2007. *30*: pp. 259–288.
149. Gopnik, A. Explanation as orgasm. *Minds and Machines,* 1998. *8*: pp. 101–118.
150. Sanfey, A. G. The neural basis of economic decision-making in the Ultimatum Game. *Science,* 2003. *300*(5626): pp. 1755–1758.

151. King-Casas, B., et al. Getting to know you: Reputation and trust in a two-person economic exchange. *Science,* 2005. *308*(5718): pp. 78–83.
152. Jacobsen, T., et al. The primacy of beauty in judging the aesthetics of objects. *Psychological Reports,* 2004. *94*: pp. 1253–1260.
153. Carroll, N. Beauty and the genealogy of art theory. *Philosophical Forum.,* 1991. 22(4): pp. 307–334.
154. Sibley, F. Aesthetic and nonaesthetic. *Philosophical Review,* 1965. 74(2): pp. 135–159.
155. Gombrich, E. H. *The Story of Art.* 1950, London: Phaidon.
156. Hutcheson, F. *An Inquiry into the Original of Our Ideas of Beauty and Virtue in Two Treatises.* 1725/2004, Indianapolis: Liberty Fund.
157. Burke, E., *A Philosophical Inquiry into the Origin of Our Ideas of the Sublime and Beautiful.* 1757/1998, New York: Oxford University Press.
158. Hume, D. *Of the Standards of Taste,* J. W. Lenz (Ed.). 1757/1965, Indianapolis: Bobbs-Merrill.
159. Kant, I., *Critique of Judgment,* W. S. Pluhar (Transl.), 1790/1987, Indianapolis: Hackett.
160. Bullough, E. "Psychical distance" as a factor in art and and aesthetic principle. *British Journal of Psychology,* 1904–1920, 1912. 5(2): pp. 87–118.
161. Bell, C., *Art,* J. B. Bullen (Ed.). 1914/1987, Oxford: Oxford University Press.
162. Shiner, L., *The Invention of Art. A Cultural History.* 2001, Chicago: University of Chicago Press.
163. Fischer, S. R. *A History of Reading.* 2003, London: Reaktion Books.
164. Bub, D. N., Arguin, M., & Lecours, A. R. Jules Dejerine and his interpretation of pure alexia. *Brain and Language,* 1993, *45,* pp. 531–559.
165. Nakamura, K., et al. Universal brain systems for recognizing word shapes and handwriting gestures during reading. *Proceedings of the National Academy of Sciences U S A,* 2012, *109*(50): pp. 20762–20767.
166. Holmes, G. Disturbances of visual orientation. *British Journal of Ophthalmology,* 1918. 2: pp. 449–468.
167. Zeki, S. Art and the brain. *Journal of Consciousness Studies,* 1999. *6*: pp. 76–96.
168. Zeki, S., & Lamb, M. The neurology of kinetic art. *Brain,* 1994. *117* (Pt 3): pp. 607–636.
169. Cavanagh, P. The artist as neuroscientist. *Nature,* 2005. *434*(7031): pp. 301–307.
170. Seeley, W. P. What is the cognitive neuroscience of art...and why should we care? *American Society of Neuroaesthetics Newsletter,* 2011. *31*(2): pp. 1–4.
171. Livingstone, M. *Vision and Art: The Biology of Seeing.* 2002, New York: Abrams.
172. Ungerleider, L. G., & Mishkin, M. Two cortical visual systems. In *Analysis of Visual Behavior,* D. J. Ingle, M. A. Goodale, & R. J. W. Mansfield (Eds.). 1982, Cambridge, MA: MIT Press, pp. 549–586.
173. Carroll, N. *Art in Three Dimensions.* 2010, New York: Oxford University Press.

174. Ekman, P. An argument for basic emotions. *Cognition & Emotion*, 1992. *6*(3–4): pp. 169–200.
175. Roseman, I., & Evdokas, A.Appraisals cause experienced emotions: Experimental evidence. *Cognition & Emotion*, 2004. *18*(1): pp. 1–28.
176. Rock, I., & Palmer, S. The legacy of Gestalt psychology. *Scientific American*, 1990. *263*(6): pp. 84–90.
177. Arnheim, R. *Art and Visual Perception: A Psychology of the Creative Eye*. 1954, Berkeley, CA: University of California Press.
178. Berlyne, D. Novelty, complexity and hedonic value. *Perception and Psychophysics*, 1970. *8*: pp. 279–286.
179. Taylor, R. P., Micolich, A. P., & Jonas, D. Fractal analysis of Pollock's drip paintings. *Nature*, 1999. *399*(6735): pp. 422–422.
180. Taylor, R. P., et al. Authenticating Pollock paintings using fractal geometry. *Pattern Recognition Letters*, 2007. *28*(6): pp. 695–702.
181. Jones-Smith, K., & Mathur, H. Fractal analysis: Revisiting Pollock's drip paintings. *Nature*, 2006. *444*(7119): pp. E9–E10.
182. Redies, C. A universal model of esthetic perception based on the sensory coding of natural stimuli. *Spatial Vision*, 2007. *27*: pp. 97–117.
183. Graham, D. J., & Field, D. J. Statistical regularities of art images and natural scenes: Spectra, sparseness and nonlinearities. *Spatial Vision*, 2007. *21*: pp. 149–164.
184. Graham, D. J., & Redies, C. Statistical regularities in art: Relations with visual coding and perception. *Vision Research*, 2010. *50*(16): pp. 1503–1509.
185. Graham, D. J., & Field, D. J. Variations in intensity statistics for representational and abstract art, and for art from the Eastern and Western hemispheres. *Perception*, 2008. *37*(9): pp. 1341–1352.
186. Redies, C., et al. Artists portray human faces with the Fourier statistics of complex natural scenes. *Network (Bristol, England)*, 2007. *18*(3): pp. 235–248.
187. Chatterjee, A. Prospects for a cognitive neuroscience of visual aesthetics. *Bulletin of Psychology and the Arts*, 2004. *4*: pp. 55–59.
188. Kawabata, H., & Zeki, S. Neural correlates of beauty. *Journal of Neurophysiology*, 2004, *91*(4): pp. 1699–705.
189. Vartanian, O., & Goel, V. Neuroanatomical correlates of aesthetic preference for paintings. *Neuroreport*, 2004. *15*(5): pp. 893–897.
190. Jacobsen, T., et al. Brain correlates of aesthetic judgments of beauty. *Neuroimage*, 2005. *29*: pp 276–285.
191. Cela-Conde, C. J., et al. The neural foundations of aesthetic appreciation. *Progress in Neurobiology*, 2011. *94*(1): pp. 39–48.
192. Biederman, I., & Vessel, E.A. Perceptual pleasure and the brain. *American Scientist*, 2006, *94*: pp. 249–255.
193. Leder, H., Carbon, C.-C., & Ripsas, A.-L. Entitling art: Influence of title information on understanding and appreciation of paintings. *Acta Psychologica*, 2006. *121*(2): pp. 176–198.

194. Jakesch, M., & Leder, H. Finding meaning in art: Preferred levels of ambiguity in art appreciation. *Quarterly Journal of Experimental Psychology*, 2009. *62*(11): pp. 2105–2112.
195. Kirk, U., et al. Modulation of aesthetic value by semantic context: An fMRI study. *Neuroimage*, 2009. *44*(3): pp. 1125–1132.
196. Cutting, J. E. Mere exposure, reproduction, and the impressionist canon. In *Partisan Canons*, A. Brzyski (Ed.). 2007, Durham, NC: Duke University Press: pp. 79–93.
197. Wiesmann, M., & Ishai, A. Training facilitates object recognition in cubist paintings. *Frontiers in Human Neuroscience*, 2010. *4*: p. 4.
198. Kirk, U., et al. Brain correlates of aesthetic expertise: A parametric fMRI study. *Brain and Cognition*, 2009. *69*(2): pp. 306–315.
199. Danto, A. C. *The Abuse of Beauty: Aesthetics and the Concept of Art. Paul Carus Lectures; 21st Series*. 2003. Chicago: Open Court Publishing.
200. Gopnik, B. Aesthetic science and artistic knowledge. In *Aesthetic Science: Connecting Minds, Brains and Experience*, A. P. Shimamura & S. E. Palmer (Eds.). 2012, New York: Oxford University Press, pp. 129–159.
201. Eskine, K. J., Kacinik, N. A., & Prinz, J. J. Stirring images: Fear, not happiness or arousal, makes art more sublime. *Emotion*, 2012. *12*(5): pp. 1071.
202. Nodine, C. F., Locher, P. J., & Krupinski, E. A. The role of formal art training on perception and aesthetic judgement of art compositions. *Leonardo*, 1993. *26*(3): pp. 219–227.
203. Buskirk, J. Artful arithmetic: Barthel Beham's Rechner and the dilemma of accuracy. *Renaissance Quarterly*, 2013.
204. McBrearty, S., & Brooks, A. S. The revolution that wasn't: A new interpretation of the origin of modern human behavior. *Journal of Human Evolution*, 2000. *39*(5): pp. 453–563.
205. Nowell, A. Defining behavioral modernity in the context of Neandertal and anatomically modern human populations. *Annual Review of Anthropology*, 2010. *39*(1): pp. 437–452.
206. Balter, M. Origins. On the origin of art and symbolism. *Science*, 2009. *323*(5915): pp. 709–711.
207. Moore, M. W., & Brumm, A. R. Symbolic revolutions and the Australian archaeological record. *Cambridge Archeological Journal*, 2005. *15*(2): pp. 157–175.
208. Henshilwood, C. S., d'Errico, F., & Watts, I. Engraved ochres from the Middle Stone Age levels at Blombos Cave, South Africa. *Journal of Human Evolution*, 2009. *57*(1): pp. 27–47.
209. Bouzouggar, A., et al. 82,000-year-old shell beads from North Africa and implications for the origins of modern human behavior. *Proceedings of the National Academy of Sciences U S A*, 2007. *104*(24): pp. 9964–9969.
210. Vanhaereny, M., et al. Middle Paleolithic shell beads in Israel and Algeria. *Science*, 2006. *312*(5781): pp. 1785–1788.

211. Barham, L. Backed tools in Middle Pleistocene central Africa and their evolutionary significance. *Journal of Human Evolution*, 2002. *43*(5): pp. 585–603.
212. Marean, C. W., et al. Early human use of marine resources and pigment in South Africa during the Middle Pleistocene. *Nature*, 2007. *449*(7164): pp. 905–908.
213. Roebroeks, W., et al. Use of red ochre by early Neandertals. *Proceedings of the National Academy of Sciences U S A*, 2012. *109*(6): pp. 1889–1894.
214. Schwarcz, H. P., & Skoflek, I. New dates for the Tata, Hungary archaeological site. *Nature*, 1982. *295*: pp. 590–591.
215. Peresani, M., et al. Late Neandertals and the intentional removal of feathers as evidenced from bird bone taphonomy at Fumane Cave 44 ky B.P., Italy. *Proceedings of the National Academy of Science U S A*, *108*: pp. 3888–3893.
216. Haidle, M. N., & Pawlik, A. F. The earliest settlement of Germany: Is there anything out there? *Quaternary International*, 2010. *223–224*: pp. 143–153.
217. d'Errico, F., & Nowell, A. A new look at the Berekhat ram figurine: Implications for the origins of symbolism. *Cambridge Archaeological Journal*, 2000, *10*(01): pp. 123–167.
218. Bednarik, R. G. A figurine from the African Acheulian. *Current Anthropology*, 2003. *44*(3): pp. 405–413.
219. Bednarik, R. G. The earliest evidence of palaeoart. *Rock Art Research*, 2003. *20*(2): pp. 89–135.
220. Edwards, S. W. Nonutilitarian activities in the Lower Paleolithic: A look at the two kinds of evidence. *Current Anthropology*, 1978. *19*(1): pp. 135–137.
221. Oakley, K. P. Emergence of higher thought 3.0-0.2 Ma B.P. *Philosophical Transactions of the Royal Society, Series B: Biological Sciences*, 1981. *292*(1057): pp. 205–211.
222. Nowell, A. From a Paleolithic art to Pleistocene visual cultures (introduction to two special issues on "Advances in the Study of Pleistocene Imagery and Symbol Use"). *Journal of Archaeological Method and Theory*, 2006. *13*(4): pp. 239–249.
223. Conkey, M. W. New approaches in the search for meaning? A review of research in "Paleolithic art." *Journal of Field Archaeology*, 1987. *14*(4): pp. 413–430.
224. Mellars, P.A. Cognition and climate: Why is Upper Palaeolithic cave art almost confined to the Franco-Cantabrian region? In *Becoming Human. Innovation in Prehistoric Material and Spiritual Culture*, C. Renfrew & I. Morley (Eds.). 2009, New York: Cambridge University Press, pp. 212–231.
225. Andrews, P. W., Gangestad, S. W., & Matthews, D. Adaptationism—how to carry out an exaptationist program. *Behavioral and Brain Sciences*, 2002. *25*(4): pp. 489–504, discussion 504–553.
226. Greene, E. A diet-induced developmental polymorphism in a caterpillar. *Science*, 1989. *243*(4891): pp. 643–646.

227. Harvell, C. D. The ecology and evolution of inducible defenses in a marine bryozoan: Cues, costs, and consequences. *American Naturalist*, 1986. *128*(6): pp. 810–823.
228. Fernald, R. D., & Hirata, N. R. Field study of *Haplochromis burtoni*: Quantitative behavioural observations. *Animal Behaviour*, 1977. 25: pp. 964–975.
229. Maruska, K. P., & Fernald, R. D. Plasticity of the reproductive axis caused by social status change in an African cichlid fish: II. Testicular gene expression and spermatogenesis. *Endocrinology*, 2011. *152*(1): pp. 291–302.
230. Pearce-Duvet, J. M. C. The origin of human pathogens: Evaluating the role of agriculture and domestic animals in the evolution of human disease. *Biological Reviews*, 2006. *81*(3): pp. 369–382.
231. Wiesenfeld, S. L. Sickle-cell trait in human biological and cultural evolution. *Science*, 1967. *157*(3793): pp. 1134–1140.
232. Gould, S. J., & Lewontin, R. C. The spandrels of San Marco and the Panglossian paradigm: A critique of the adaptationist programme. *Proceedings of the Royal Society of London. Series B, Biological Sciences.*, 1979. *205*(1161): p. 581–598.
233. Dissanayake, E. The arts after Darwin: Does art have an origin and adaptive function? In *World Art Studies: Exploring Concepts and Approaches*, K. Zijlemans & W. van Damme (Eds.). 2008, Amsterdam: Valiz, pp. 241–263.
234. Dissanayake, E. *Art and Intimacy. How the Arts Began*. 2000, Seattle: University of Washington Press.
235. Miller, G., *The Mating Mind: How Sexual Choice Shaped the Evolution of Human Nature*. 2000, New York: Doubleday.
236. Snell-Rood, E. C., et al. Toward a population genetic framework of developmental evolution: The costs, limits, and consequences of phenotypic plasticity. *BioEssays*, 2010. *32*(1): pp. 71–81.
237. Deacon, T. W. Colloquium paper: A role for relaxed selection in the evolution of the language capacity. *Proceedings of the National Academy of Sciences U S A*, 2010. *107*(Suppl 2): pp. 9000–9006.
238. Okanoya, K. The Bengalese finch: A window on the behavioral neurobiology of birdsong syntax. *Annals of the New York Academy of Sciences*, 2004. *1016*(1): pp. 724–735.
239. Yamazaki, Y., et al. Sequential learning and rule abstraction in Bengalese finches. *Animal Cognition*, 2011. *15*(3): pp. 369–377.
240. Hosino, T., & Okanoya, K. Lesion of a higher-order song nucleus disrupts phrase level complexity in Bengalese finches. *Neuroreport*, 2000. *11*(10): pp. 2091–2095.
241. Rampen, J., & Tuffrey, L. How Arab revolutionary art helped break the spell of political oppression. *The Guardian* 2012. http://www.guardian.co.uk.
242. AFP. *China Artist Ai Weiwei Sets up Home Webcams*. 2012 Retrieved April 2, 2012, from http://www.google.com/hostednews/afp/article/ALeqM5jLfw_i27vVoxdFZZUra3srx6k3tw?docId=CNG.d33392c87708cd1e02226d7870c3a573.461.

INDEX

CPSIA information can be obtained
at www.ICGtesting.com
Printed in the USA
BVOW02s2105161116
468124BV00001B/3/P